Introduction to the
PHYSICS OF
COMPLEX SYSTEMS

*The mesoscopic approach to
fluctuations, non linearity and
self-organization*

Related Pergamon Titles of Interest

Books
A. I. AKHIEZER & S. V. PELETMINSKII
Methods of Statistical Physics

M. G. BOWLER
Lectures on Special Relativity

M. G. BOWLER
Lectures on Statistical Mechanics

V. L. GINZBURG
Physics and Astrophysics (A selection of key problems)

L. D. LANDAU & E. M. LIFSHITZ
Course of Theoretical Physics
Volume 5: Statistical Physics Part 1, 3rd edition
Volume 9: Statistical Physics Part 2

R. K. PATHRIA
Statistical Mechanics

D. TER HAAR
Collected Papers of P L Kapitza, Volume 4

Journals
Journal of the Franklin Institute

Nonlinear Analysis (Theory, Methods & Applications)

Reports on Mathematical Physics

Full details of all Pergamon publications and a free specimen copy of any
Pergamon journal available on request from your nearest Pergamon office.

Introduction to the
PHYSICS OF
COMPLEX SYSTEMS

The mesoscopic approach to fluctuations, non linearity and self-organization

by

ROBERTO SERRA, MASSIMO ANDRETTA,
MARIO COMPIANI
Tema SpA, Bologna, Italy

and

GIANNI ZANARINI
University of Bologna, Italy

PERGAMON PRESS

OXFORD · NEW YORK · BEIJING · FRANKFURT
SÃO PAULO · SYDNEY · TOKYO · TORONTO

U.K.	Pergamon Press, Headington Hill Hall, Oxford OX3 0BW, England
U.S.A.	Pergamon Press, Maxwell House, Fairview Park, Elmsford, New York 10523, U.S.A.
PEOPLE'S REPUBLIC OF CHINA	Pergamon Press, Qianmen Hotel, Beijing, People's Republic of China
FEDERAL REPUBLIC OF GERMANY	Pergamon Press, Hammerweg 6, D-6242 Kronberg, Federal Republic of Germany
BRAZIL	Pergamon Editora, Rua Eça de Queiros, 346, CEP 04011, São Paulo, Brazil
AUSTRALIA	Pergamon Press Australia, P.O. Box 544, Potts Point, N.S.W. 2011, Australia
JAPAN	Pergamon Press, 8th Floor, Matsuoka Central Building, 1-7-1 Nishishinjuku, Shinjuku-ku, Tokyo 160, Japan
CANADA	Pergamon Press Canada, Suite 104, 150 Consumers Road, Willowdale, Ontario M2J 1P9, Canada

First edition 1986

Library of Congress Cataloging in Publication Data

Introduzione alle fisica dei sistemi complessi. English.
Introduction to the physics of complex systems.
Translation of: Introduzione alle fisica dei sistemi complessi.
Bibliography: p.
Includes index.
1. Fluctuations (Physics)   2. Nonlinear theories.
3. Self-organizing systems.   4. System theory.
5. Stochastic processes.   I. Serra, Roberto.   II. Title.
QC6.4.F58I5813   1986   530.1   86-12321

British Library Cataloguing in Publication Data

Introduction to the physics of complex systems : the mesoscopic approach to fluctuations, non linearity and self-organization.
1. System theory
I. Serra, Roberto   II. Introduzione alle fisica dei sistemi complessi. *English*
003 Q295
ISBN 0-08-032629-3 (Hardcover)
ISBN 0-08-032628-5 (Flexicover)

A translation of *Introduzione alle Fisica dei Sistemi Complessi* (L'approcio mesoscopico allo studio di fluttuazioni, non linearita e auto-organizzazione) published by Editrice CLUEB Bologna.
The camera ready version of this book was prepared by Studio T-Line Bologna (Italy).

Printed in Great Britain by A. Wheaton & Co. Ltd., Exeter

To Elena, Marlena,
Silvia and Giovanna

Foreword

Let me start with an elementary outlook on what is nowadays called the physics of complex systems, and on its connections with more traditional aspects of classical physics, such as Newtonian mechanics and thermodynamics.

Until recently, chaos, order and complexity were regarded as three different levels of description of physical reality. Since they were considered in different historical periods, they seemed related to different cultural backgrounds, that is, dependent on the particular class of phenomena considered to be relevant by the observer. This led to some degree of arbitrariness in the physicist's approach to nature, giving rise to the suspicion that the description was biased by his philosophical background. On the contrary, it will be shown here that the the three levels may coexist in the same physical system, corresponding to different degrees of exchange with the outside world.

The first men who raised scientific questions about the world were obliged to make drastic choices among a manifold of equally interesting facts. The philosophical and poetic attempts to transfer within the confines of a written language the "totality" of a piece of reality led to disputes from the beginning. One may recall the conflict between "being" and "becoming" in the pre-Aristotelian schools and the attempts at synthesis by Aristotle. Even when investigators were not fostering ambitions of "wholeness", as with Arabian or Renaissance scientists, the scientific product often seemed to be a catalogue of events with only empirical connections without logical correlations.

On the other hand some men had enough wisdom (and creativity) to give up such ambitions of "wholeness" and limit the description to some evident regularities. Such was Galileo's and Newton's revolution when they isolated from the whole reality simple items such as the pendulum or the falling stone. Thus, the physics of Galileo and Newton are based upon two models of natural behaviour: the pendulum and the two-body gravitational problem.

Let me list the main elements of such a revolutionary approach. First, isolating a few quantities, after the observation these can be treated as numbers and hence related mathematically, within the framework of a suitable calculus. Second, these few elements can be organised into a simple experiment which can be verified anywhere. The criterion of truth, in this limited realm of knowledge, does not imply a philosophy, but is related only to the experimental verification.

ORDER

The success of this reductive program in explaining regularities fostered the presumption that all phenomena would be described as the resultant of local motions of elementary constituents (mechanism). The passage from 2 to N bodies (where N can be as large as the Avogadro number) was considered legitimate in principle, even though the only rigorously solved problem was the two-body one. This dogma influenced not only the conceptual construction of the solar system, but also that of atoms, molecules, and nuclei.

All the physics springing from mechanism may be considered as a static machine, obtained by adding together components in dynamic equilibrium (i.e. with mutual forces compensated by reactions), and then superposing varieties of known motions upon the static machine. Such is the physical description of a crystal, and such is also the description of any technological device, since rotational or alternating motions are particular cases of the pendulum.

This physics, born from a drastic simplification of natural complexity, was assigned the task of describing "order". It pretended to make predictions by considering the "whole" as a sheer sum of elementary components. On the contrary, we know today that a many-body system can behave in ways which can not be predicted from elementary two-body interactions.

Furthermore this physics of "order" is affected by a kind of unreality. In its equations time can legitimately flow in the reverse direction. If one changes the sign of the time variable, we obtain fully symmetric solutions. For three hundred years it has been current opinion that time irreversibility is a subjective outcome of our limited resolution during observations, but that "true" reality is described by reversible equations. Now, such a symmetry does not hold in nature, apart from the very simple systems mentioned above, and we must introduce irreversibility at the very beginning of our logical journey.

STATISTICAL CHAOS, OR DISORDER

During the last century investigations into thermal engines, i.e. thermodynamics, added a new chapter to physics where a new concept was introduced, namely entropy, or the degree of disorder of a multicomponent system. When a system is in thermal equilibrium, that is, at a uniform temperature, it has reached a situation of maximum disorder, or chaos. A new physical law, the second principle of thermodynamics, states that an isolated system must evolve toward equilibrium, therefore, entropy must grow toward a maximum. This law corresponds to the introduction of quality in a universe of quantities. It puts an "arrow" to time, saying that nature evolves in a unique direction and cannot move backwards.

This is indeed a common experience in everyday life, and being unable to account for this was the main limitation of the physics of order. This qualitative law introduced a conceptual gap, not yet filled, into the body of physics. The attempts to elaborate a new mechanics (statistical mechanics) necessarily imply a non–mechanistic assumption in order to account for the entropy law, such as Boltzmann's "Stosszahlansatz".

DETERMINISTIC CHAOS, OR TURBULENCE

The third level of description corresponds to the dynamics born with the three-body problem in celestial mechanics. Already for N = 3 a dynamic system is very different from the two-body problem, since in general there are asymptotic instabilities. This means a divergence, exponential in time, of two trajectories in phase space stemming from initially close points. The uniqueness theorem for solutions of differential systems seems to offer a means of escape: to be more and more precise in localising the coordinates of the initial point. However a fundamental difficulty arises. Only rational numbers can be assigned by a finite number of digits. A "precise" assignment of a real number requires an infinite acquisition time and an infinite memory capacity to store it, and neither of these two infinities is available to the physicist. Hence any initial condition implies a truncation. A whole range of initial conditions, even if small, is usually given and from within it trajectories may arise whose difference becomes sizeable after a given time, if there is an exponential divergence. This means that predictions are in general limited in time and that motions are complex, even starting from the three-body case. In fact we now know from very elementary topological considerations that a three-dimensional phase space corresponding to three coupled degrees of freedom is already sufficient to yield a positive Lyapunov exponent, and accordingly an expanding phase space direction. This complexity is not due to coupling with a noise source such as a thermal reservoir, but to inavoidable inaccuracy in setting the initial conditions. It is called deterministic chaos.

The birth of this new dynamics was motivated by practical problems, such as fixing the orbit of a meteorological satellite, and it was strongly helped by the introduction of powerful computers. The mathematics of multiple bifurcations leading from a simple behaviour to a complex one is the subject of current investigation. Some regularities, such as the so called "scenarios" or routes to deterministic chaos, appear as exciting and mysterious as the old spectroscopic rules before their organisation in the quantum formalism. We are hopefully on the verge of a new formalism, which could give an integral description of the passage from order to complicated behaviours such as fully-developed turbulence in a fluid.

This three-body (or several-body) dynamics leading to deterministic chaos is the natural extension of the two-body, or Newtonian dynamics without, however, the ideological implications associated with full predictability, or determinism.

COMPLEXITY

When we consider a many-body system made up of many interacting components, three cases are conceptually possible.

i)   The mutual interactions can be linearised. A linear system can be diagonalised, that is, transformed into a set of uncoupled degrees of freedom, or normal modes, and hence we face just a one-body dynamics (e.g. the phonon theory).

ii) The nonlinearities are essential, and we are in the presence of many coupled degrees of freedom. There may be however a heuristic rule to distinguish between a "system" which is the part we are interested in and a "thermal bath" which is the rest of the world weakly perturbing the system. We have to use a statistical approach, and the fluctuation-dissipation theorem tells us that the effect of the bath is to provide a deterministic dissipation plus noise. The system may have a reasonable, but still large, number of degrees of freedom (3 in a single mode laser, 10^6 in a cubic centimeter of stirred water undergoing a turbulent motion). If all of them are equally "fast" (that is, have comparable damping times) then one has to expect deterministic chaos, and only recently developed dynamics is able to tackle some qualitative aspects of their motion.

iii) In the latter case it may happen that one (or a few) degrees of freedom are particularly slow (such as the tuned mode in a high Q electromagnetic cavity). On these long time scales the fast variables can be considered at equilibrium (adiabatic elimination) and the whole dynamics becomes low-dimensional, sometimes even one-dimensional. In contrast with case i), however, the resulting dynamics is highly nonlinear and multiple bifurcations may occur depending upon the values of the "control parameters" which summarise the effect of the fast variables. Some of these bifurcations are the same as those already considered in thermodynamic phase transitions and may lead to highly ordered states starting from a disordered phase. In this way, we recover the characteristic of order which was once considered peculiar to one and two-body problems.

Let us now see how disorder, order and deterministic chaos play a role in so-called "open systems". We use this term for a many-body system with nonlinear interactions among its components and away from thermal equilibrium. It is "complex" in the sense that the same individual object can display all three behaviours (order, disorder and deterministic chaos or turbulence).

OPEN SYSTEMS: TWO EXAMPLES

Equilibrium statistical mechanics, or its linearised nonequilibrium version, deal with phenomena arising from the coupling of a system with a thermal reservoir. Much more interesting is the physics of "open systems", that is, of bodies which transfer energy from a source to a sink. Source and sink can be large reservoirs at constant temperature; however, the body can not be described in terms of a uniform temperature. This is for instance the case of a living being which receives nutrients and disposes of metabolic residues. Depending upon the values of the "control parameters", the same open system can go from chaos to order and then to turbulence. Hence, the three levels no longer appear as three possible conceptual abstractions, but as three different behaviours of the same physical system.

Let us consider the laser. The laser is a light source consisting of a tube filled with luminescent atoms or molecules with a mirror at each end to reflect the light, ensuring prolonged "contact" with the atoms. As the energy supplied to the atoms is increased, there are three successive situations, namely:

a) DISORDER - For low excitation, we have a normal light source which emits light in all directions and of many colours. The photoelectric output from a fast detector appears as a noisy signal without apparent correlations.

b) ORDER - Increasing the excitation above a threshold value the system undergoes "symmetry breaking". Only one direction and one colour are favoured, all the others being quenched. The photoelectric signal is highly ordered.

How did order spring from chaos? We sketch here a heuristic answer. In thermal equilibrium, any component of a large system executes a small motion described by linear dynamics (like the harmonic oscillator or the linearised pendulum). Such small motions are necessary to balance that which each component obtains from elsewhere and what it has to give back, in order to keep a constant temperature. In such small motions the component explores only its immediate surroundings and no long range features can build up in the system. For larger motions, the components must necessarily interfere with one another's behaviour, and beyond a threshold the motion becomes "cooperative", that is, ordered over a long range. In the laser case, as the field increases beyond a critical value, the excited molecules contribute by "stimulated emission" to build up the same "coherent"field, rather than emitting uncorrelated contributions.

c) TURBULENCE - The onset of order was explained in terms of all radiators driven by a single field which imposes its phase. Technically, this corresponds to a dynamics with a single relevant degree of freedom (the field), all the others being "slaved" to adapt instantly to the slowly varying main variable. We have already exploited this "slaving" in the adiabatic elimination of the fast variables. When, however, the excitation increases above a second threshold, all the dynamic variables must be equally considered. The single mode laser happens to be described by just three collective variables: the electromagnetic field, the induced polarization and the population. But three is a sufficient number for having a positive Lyapunov exponent, and hence an expanding direction in phase space. Thus deterministic chaos may arise.

The Rayleigh-Bénard hydrodynamic instability may be described in a similar manner. The physical system consists of a cell with rigid walls filled with a liquid and heated from below. As the temperature difference between bottom and top increases, we have three distinct regimes, namely:

a) DISORDER - Initially the molecules are heated at the bottom, that is, they receive an excess of kinetic energy with respect to the average, and they transfer their excitation by individual molecular collisions. This process is heat conduction and no particular correlations appear with respect to the isothermal fluid.

b) ORDER - Above a critical temperature difference, convective motions start. Small regions of fluid near the bottom acquire a lower density and rather than rapidly thermalising with their surroundings, behave as minute "bubbles" pushed up by buoyancy. This motion is correlated over the whole cell, and can be visualised by ink droplets, showing a "roll" feature, since the upward motion has to turn anyway at the top where the constraint of the wall stops the upward tendency.

c) TURBULENCE - On further heating, more degrees of freedom play a relevant role, thus expanding directions in phase space arise, with associated irregularities in the motion. The overall motion appears as turbulent.

A DETERMINISTIC VS. STOCHASTIC DESCRIPTION OF TRANSITIONS

We have thus seen that open systems can successively undergo transitions, passing through configurations which were once considered as mutually exclusive and fully characterising a given body.

In the above examples the dynamics was described as the competition of two antagonists, namely a loss mechanism (the light leakage in the laser case, the molecular collisions generating viscosity in the hydrodynamic case) and a gain mechanism tending to induce long-range dynamics. This was the case of stimulated emission for the laser, and buoyancy for the Rayleigh-Bénard convection.

We have called "order" the appearance of time and space regularities, or patterns, for which the description in terms of a few variables becomes "economically fruitful", yielding a high predictability.

Above we have described the deterministic transition to order when gains overcome losses. In mathematical terms we have a bifurcation where an unstable branch disappears and a new stable branch arises "spontaneously". However, in nature there is no "spontaneous" birth. Indeed an open system is exposed to many interactions with the external world, whose gross features are summarised by a dynamic law as if the system were isolated. Also, there is an unpredictable behaviour, which can only be described statistically. The position of the system in phase space at time t, called $X(t)$, will be obtained by the application of an evolution operator $A(t, t')$ to the position $X(t')$ at a previous time t'. This is the deterministic evolution. Furthermore there will be a stochastic contribution, or noise, that we denote by ξ. The physical law will be:

$$X(t) = A(t, t') X(t') + \xi$$

A typical example is Brownian motion. Even when $A = 1$, the resultant of all molecular collisions upon the tiny pollen bodies will still induce an extra, chaotic, motion.

Noise is usually negligible, because its relative effect, with respect to the deterministic one, scales with the reciprocal of the square root of the total number N of particles in the system. Hence for large systems this effect is usually small and below the experimental resolution. However, the system can go through critical points or bifurcations where the macroscopic thrust A falls temporarily to zero and hence the role of noise becomes crucial.

ABOUT THIS BOOK

The intellectual challenge offered by the above considerations has to be transformed into working tools for a detailed interpretation of physical phenomena. Perhaps, in the future, when these arguments will be well established in the science curricula, the approach to them will be radically different than today. I expect in the future a more extensive use of discrete tools, related to the mathematical processing on computers, with a corresponding de-emphasis of the role of the continuum approach. At present, the amount of knowledge needed by the average scientist for an active investigation of complexity is just that offered in this book by the Authors. The sequence of subjects they have chosen in their presentation is the best interface possible between the actual cultural background of a physicist, or chemist, or electrical engineer, and this new field.

We have seen that the area denoted by the key words "fluctuations, nonlinearity and self-organisation" is a common ground to both nonequilibrium statistical mechanics and system dynamics. Haken has introduced the word "synergetics" for it. Very few books are available in this overlap are which is growing as a new discipline. They are included in the Bibliography of this volume. However I must say that all of them are unbalanced in some way. Since they reflect the specific working experience of the writers, they over-emphasize some detailed technique and give an insufficient coverage of other useful items.

On the contrary, the present Authors, having entered the field of complex systems from different backgrounds, had to face by themselves the task of mastering all the useful tools, and hence took great care in providing whatever seemed necessary to deal with these problems.

The first four of the six chapters cover nonequilibrium statistical mechanics, the fifth chapter deals with system dynamics, and the last one offers an outline of synergetic systems. The Authors provide an extremely clear but rigorous presentation. The book, as it is, is the most useful Introduction which can be provided to a graduate student or a young researcher entering this field. However, the reader may then realise that one single chapter on system dynamics may not be sufficient, and hence he will refer to excessively specialised books such as the classic Minorski or the more recent Guckenheimer and Holmes. Let us hope that the Authors will soon be able to offer us an equally clear and extensive approach to system dynamics, as they have here provided for stochastic processes, in a second book.

Florence, January 1986   *F.T. ARECCHI*

Preface

It is an extremely trying, although fascinating, experience to actively reconsider one's own professional identity and, at the same time, one's own way of looking at the world. Naturally, this task would not be carried out without a valid reason: in our case, it came from the impression that the basic approach towards physical reality was gradually changing, and that something really new and different was happening within physical science.

Something which, in order to be understood and become our own, required us to reorder all that we were learning about nonlinear systems and stochastic processes, to put together concepts which, traditionally, were remote from each other and to develop a "mesoscopic approach" towards physical phenomena.

As we had hoped, this allowed us to begin to understand unfamiliar and counterintuitive behaviour of complex systems, while rediscovering all the appeal and the richness of classical physics.

Developing a "science of complexity" for ourselves, in the absence of systematic texts, was a fairly lengthy and tiring exercise: we hope that the reader of the present book will save some of this time and effort. It is clear that, for us, this organisation of the subject (partial, debatable, dated, as in all such cases) is not an end in itself; it should rather be considered as an introduction to the activity of research in the field of complex systems. This will, we think, be an assurance for those who, in this book, are not so much looking for an absolutely formal and rigorous treatment, but rather for a "launching pad" for further activity.

The preparative study and the drafting of this book would not have been possible without the help of friends and colleagues who, being ahead of us in specific fields, helped us in overcoming the difficulties which occasionally arose. A special acknowledgement is due to Bernard H. Lavenda who introduced us to the statistical thermodynamics of irreversible processes, and to Paolo Grigolini, who gave a decisive contribution to our understanding of the theory of stochastic processes. We are specially indebted to F. Tito Arecchi, one of the leading experts in laser physics and deterministic chaos, for his scientific support and for his kind foreword to this work.

We would also like to thank Rita Casadio for the preparation of the experiment described in the Appendix.

Particular acknowledgements are due to our friend and colleague Laura Gardini whose intimate knowledge of the theory of dynamic systems was of great help to us during the drafting of the last two chapters.

The development of this work was made possible by the financial and cultural support of

ENI (contract Nr. 3235/24/2/1982), ENIDATA (contracts Nrs. 485ED84 and 020ED86) and TEMA, the company where three of us (RS, MA, MC) work. In particular, we would like to thank Vincenzo Gervasio, Paolo Verrecchia and Silvio Serbassi for their advice and encouragement.

We would also like to thank M.W. Evans, K. Pendergast and M. Maestro who read the work and helped us with precious suggestions.

The present book is an expanded version of a previous one, "Introduzione alla Fisica dei Sistemi Complessi", CLUEB, Bologna, 1984. The translation was carried out by Dr. D. Jones, whom we thank for his excellent work.

The preparation of the text would not have been possible without the perseverance and professionality of Anna Lodi and Giulia Bariselli to whom special thanks are due.

Finally we would like to thank Peter Henn, Senior Publishing Manager of Pergamon Press Ltd., together with his staff, for their patience, competence and for the continued interest with which they have followed our work.

Bologna, February 1986   *Roberto Serra*
                                                                Gianni Zanarini
                                                                Massimo Andretta
                                                                Mario Compiani

Contents

"Right at the start we have to make a statement which you are certainly familiar with, but it is necessary to repeat it again and again. It is that science does not try to explain, nor searches for interpretations but primarily constructs models. A model is a mathematical construction, which supplemented with some verbal explanation, describes the observed phenomena. Such a mathematical construction is proved if and only if it works, that is it describes precisely a wide range of phenomena. Furthermore it has to satisfy certain aesthetic criteria, i.e. it has to be more or less simple compared to the described phenomena".

J. Von Neumann

It is impossible to know all the causal factors. We can only choose a limited number of significant factors, and use these for predicting future events, being forced to ignore factors having only a minor influence.
The hypothesis that certain events are causally connected must be accompanied by a hypothesis about the probable effects of the ignored factors.

H. Reichenbach

Everything that happens in our world resembles a vast game in which nothing is determined in advance but the rules, and only the rules are open to objective understanding. The game itself is not identical with either its rules or with the sequence of chance happenings that determine the course of play. It is neither the one nor the other because it is both at once. It has as many aspects as we project onto it in the form of questions.

M. Eigen R. Winkler

Introduction

Few subjects are as fascinating as the problems of order and disorder in physical systems, of the transitions between one condition and the other, of the intimate relationship between order and disorder in the determination of the behaviour of complex systems. Curiously enough, these subjects have never been central to classical physics.

On the other hand, it is known that the space-time scales used to describe and to understand phenomena critically determine that which one sees and that which one does not see. In particular, the framework of the analysis must be chosen with great care (or be found by a happy co-incidence, as we shall see for the case of Brownian motion) so that the interaction between order and disorder can emerge in a dynamic and constructive manner.

In the course of the treatment which follows, we will quantitatively develop this theme. For the moment, it is important to provide some aspects which will be necessary for an intuitive understanding.

First of all, within classical physics there is an indication of a general nature for choosing the "point of view" for the description of physical systems.

Within this paradigm, a "good" point of view for a given level of description is that which individuates the suitable fundamental elements (i.e. the "atoms", in a general sense). On this basis, the reality being studied may be "explained" without taking into account phenomena of an inferior level, that is, the dynamics within the atoms themselves.

It is clear that the nature of the "atoms" depends upon the context: they could be, for example, galaxies for those who study the dynamics of the universe, planets and stars for those interested in the dynamics of solar systems, or polyatomic units in the study of polymer conformation.

The choice of atoms, as mentioned above, fixes the lower limit for the level of detail of the analysis, discriminating between phenomena which are relevant or irrelevant ("sub-atomic").

Every time it happens that this process of "explaining" in terms of "atoms" proves inadequate, classical physics systematically proposes further, more detailed studies, which will lead to the identification of new atoms and to the shifting of the lower limit of relevance of the phenomena.

This descent towards the microscopic world reflects the belief, typical of classical physics, that the behaviour of every system can be completely explained, in principle, in terms of simple elements.

Within this framework, all disorder (intended as random, unpredictable or chaotic behav-

1

iour) may also be understood by moving to a lower order.

The "atomistic" tendency of resolving physical problems by using progressively smaller space-time scales may be interpreted as a continuous search for that which does not change underneath the change, for that which is reversible beneath that which appears to be irreversible.

On the other hand, the encounter with irreversibility and with the transition from order to disorder is ever-present in our daily experience. We can consider, for example, a drop of ink diffusing in water, the heat which passes from a warm body to a cold one, a gas which tends to fill all the space available to it. The reverse processes never take place spontaneously, but are only possible through an external intervention, passing, in a certain sense, from the scientific to the engineering world.

The deterministic view of classical mechanics has undergone drastic changes due to the development of quantum mechanics and the physics of elementary particles. We now know that the parts are not necessarily more simple than the whole and that, in any case, there is an insurmountable limit to our knowledge.

This "defeat" of the ambitious programme of classical physics, however, presents positive aspects of an extraordinary importance.

First of all, it allows a reduction of the caesura, of the scission which traditionally existed between science and experience, a scission based upon the essential difference between the "times" of each of these two worlds: the irreversible time of daily biological and psychic experience, the ideally reversible time of science.

Science can now look at reality in a novel manner, and in this way realise the fundamental importance (alongside decay and irreversibility) of the processes of organisation, of the creation of a localised order starting from disorder.

If, for example, we examine, within a suitable framework, even a simple disorder-to-order transition, such as the water-ice phase transition, we can observe certain fundamental aspects which are usually overlooked in the description of these phenomena.

It can be observed, in fact, that the external intervention, while being necessary for the transition to occur, has, in itself, an extremely limited information content. It is not, in fact, the reduction in temperature which provides the water with the necessary information to transform itself into ice.

This apparently trivial observation is of quite general importance. As we shall see in all of the examples of complex organised systems, from the laser to chemical oscillators, the "opening" of these systems (which carries with it a possible flow of matter or of energy) has the function of sustaining an organisation which is, fundamentally, a self-organisation.

One then realizes that, under certain conditions, "dissipative structures", to use Prigogine's appropriate term, may develop, with the occurrence of often surprising and spectacular self-organisation phenomena: complexity appears, upon which we shall concentrate our attention.

From what has been said so far, it is possible to obtain a sufficiently general definition of "complex system" which can be used as a conceptual reference for all the examples which follow. We can thus state that a "complex system"

−   is made up of a large number of mutually related elements
−   is not isolated, but has an exchange relationship (matter or energy) with the external environment
−   shows organisational characteristics which are unpredictable when considering only the individual parts of the system but which, at the same time, are not prescribed by the external environment.

As mentioned, one thus has, in complex systems, a kind of "self-organisation" or "spontaneous organisation" which "emerges" from the assembly of their different parts and from the contact with their external environment.

In order to clarify these concepts, we will refer to a particularly fascinating physical example, that of the laser, which will be dealt with in greater detail in Chapter 6.

The most obvious organisational aspect of this complex system is the coherence of the emitted radiation. In order to better illustrate this concept, we will use a particularly simple model. Let us imagine the atoms of the active material as oscillators, each of which may emit radiation. Coherence occurs when these oscillators are synchronised, "operating together". Indeed, the physicist H. Haken proposed the definition of "synergetics" for the science of complexity.

This order, this organisation is not imposed by the external environment, nor do the individual atoms "know" about it: so how can one understand its emergence? We can note that it depends upon five fundamental conditions:

— the presence of atoms of the active material

— the existence of an interaction between the atoms, mediated by the radiation present in the cavity. The radiation, produced by the oscillators, acts upon them through the quantum process of stimulated emission

— the existence of a suitable source of energy (external "pumping", which excites the oscillators)

— the existence of a particular cavity structure (shape, dimensions, presence of mirrors, etc.) which defines the boundary conditions and functions as an "ordering principle", as a "suggestion" for a certain mode of organisation

— the presence of an incessant "exploratory" activity of the possible configurations due to fluctuations.

This last point is of particular significance amongst the conditions for self-organisation, because it indicates the essential, and at the same time unexpected, importance of noise, of disorganisation, of disorder in the creation of coherence, organisation and order.

In the case of the laser, in particular, an increase in the level of external energy "pumping" leads to a bifurcation between two different possible futures (coherent emission and incoherent emission). At this point, it would seem that the determinism of the laws of physics had left space for chance, which "verifies" the effective stability or instability of a stationary state.

Many other examples of self-organisation could be described, even within physics: so many, in fact, that our way of looking at the world is changing significantly, and now, as mentioned above, we pay great attention to and marvel at realities which, in the past, we overlooked or considered to be of little significance.

Let us mention just one other case, which will help us to appreciate the role played by fluctuations: that of a fluid close to its boiling point. It is known that, under the action of a thermal gradient, produced and maintained externally, convection currents are produced within the liquid.

The remarkable aspect is that, under suitable experimental conditions (the "boundary conditions" mentioned in the discussion of the laser), these movements give rise to regular patterns ("Bénard cells" or "rolls"), which are necessarily produced by a high degree of molecular cooperation.

Here also, a random fluctuation is amplified in a non-linear manner until it gives rise to macroscopic order. Under suitable experimental conditions, the regular motion of the fluid gives rise to cylindrical structures (Bénard rolls): the warmer fluid, at the bottom, rises and tends to cool, thus descending again and being reheated, and so on. Whether the motion of the fluid is clockwise or counter-clockwise, with respect to a particular observer, is unpredictable, as the experimental apparatus is completely symmetric with respect to the two alternatives. The "decision" depends upon the random fluctuation which first occurs at the critical moment, when the previous disordered convection pattern is destabilised by the increase in the thermal gradient between the lower and the upper plates of the container, imposing a particular ordered structure.

Thus one can see the importance of the constructive relationship between complexity and chaos, between order and disorder which was overlooked by classical physics. From the examples cited above, the hypothesis which emerges is that it is the disorder itself which essentially contributes to the creation of organisation and order.

The dynamics of a transition to an organised structure, and particularly its apparently paradoxical dependence upon the presence of fluctuations, requires a novel type of description. This can be defined as a "mesoscopic" description, in order to underline the fact that it must be collocated at a suitable intermediate level which can include both order and disorder, "necessity" and "chance", using Monod's terms.

A "mesoscopic" description of complex systems must be developed in terms of new "intermediate" variables, which we will call "order parameters". The latter, as their name suggests, must be capable of describing order and organisation.

The expression "order parameters" immediately recalls the theory of Landau for equilibrium phase transitions. One of the aims, in fact, of a "science of complexity" is precisely that of extending the concepts and the mathematical techniques developed in the simple case of equilibrium to systems which are far from thermal equilibrium.

As will be seen, it is remarkable that in many cases a satisfactory "mesoscopic" description can be developed in terms of an extremely limited number of order parameters. Only a few variables, in fact, become unstable and thus play a significant role in the transitions, whereas the remaining variables may be ignored, being essentially stable and having relatively fast dynamics.

In order to clarify the idea of a mesoscopic description, we will refer to an historically important case: Brownian motion.

Let us consider a fluid containing a moving body which has a mass much larger than that of the individual fluid molecules. In this case, the effect of the presence of the fluid on the motion of the body considered gives rise to a viscous force proportional to the velocity, which transforms the constantly accelerating motion into a uniform motion. We thus have a perfectly deterministic description of the motion.

Let us now consider the case where the moving body is small enough to be affected by the fluctuations of the forces exerted by the fluid molecules. Classical mechanics still ensures the predictability of its motion, even though for practical reasons it is not possible to provide a simple description. Indeed, in this case the motion would appear erratic and unpredictable to a careful observer, and he might be tempted to conclude that such a motion should be essentially due to chance and randomness.

In the light of these considerations, we can understand the reason why the incessant and irregular motion of pollen grains in water, observed by the English botanist R. Brown in 1827, became an insoluble problem for the rest of the 19th century. The simple, but all too regular, mechanism of convection was excluded, and it was even suggested that the motion was the result of "vital" forces.

It was Einstein's ingenuity, in 1905, which suggested a treatment of phenomena of this type by combining determinism and randomness, knowledge and ignorance. He considered, in fact, a "dynamic equilibrium" between viscous force (the only force present in the case of a large object) and osmotic pressure (the only one present in the case of a small object).

Einstein's treatment of Brownian motion, later developed by Langevin in a particularly suitable form, is of fundamental importance because it has stimulated the development of the mathematical theory of stochastic processes. From our point of view, however, its importance is even greater, because it can be considered as the first example of the mesoscopic description of physical processes.

Certainly, in the case of Brownian motion itself, we can not expect to meet, as such, the phenomena of transitions and self-organisation which characterize more complex physical systems. As we shall see, however, the intuition of Einstein and Langevin with regard to the possibility and the necessity of a "mesoscopic" description is a starting point for a theory of complex physical systems.

The originality of this solution lies in the acceptance of the irreducible complexity present, and in the finding of formal instruments which would allow an acceptable representation of that complexity (the theory of stochastic processes), at the same time renouncing any further division of the system (i.e., the reduction of the problem to atomic and molecular motion).

Whether this approach may be extended to non-physical systems, and to what extent, is the subject of much debate and research at present. What we can say with certainty, is that it is quite widely accepted as a paradigm, as a conceptual framework for understanding biological, social and psychological phenomena.

The scope of the present work can not include a presentation of this debate nor of the attempts to apply the approach in many different fields. The fundamental aim here is rather that of providing (to physicists, above all) a solid and comprehensible basis for the acquisition of new working instruments, a kind of "launching pad" for novel theoretical developments and applications.

In line with what has been said so far, Brownian motion will be used in the first Chapter as the basis for the development of the fundamental concepts of the theory of "Markovian" stochastic processes, which very rapidly "forget" the past. Chapter 1 introduces the time evolution equation for the probability density (Fokker-Planck equation) and the "master equation", which allows direct treatment of systems composed of subsets of identical elements.

In the second Chapter, the concepts introduced in the first one are applied to a variety of examples, taken from physics, chemistry and electronics.

The third Chapter deals with the problem of using the previous approach for systems without apparent mesoscopic detectors such as Brownian particles. The Chapter discusses the theory of non-Markovian stochastic processes, i.e. processes having a longer "memory" than those considered previously.

In the fourth Chapter, there is a brief overview of the fundamental concepts of the thermodynamics of irreversible processes, followed by the development of its links with the theory of stochastic processes.

The fifth Chapter deals with the dynamics of the order parameters, with particular emphasis on the non-linear deterministic aspect of their evolution, which in many cases turns out to be dominant with respect to the stochastic dynamics. The aspects of stability, bifurcation and transitions which illustrate the essential role of non-linearity are discussed.

In the sixth Chapter, some sufficiently complex examples of self-organisation are presented, covering most of the concepts discussed previously.

Finally, by following the procedure given in the Appendix, it will be possible for the reader to come into direct contact with the world of complex systems, and not only with its models.

The order of reading, implicit in the arrangement of the Chapters, is not the only one possible, nor is it necessarily the best. It is simply the closest to the mental process followed by the authors in becoming acquainted with the subject of complexity.

For a first reading, one may pass directly from Chapter 2 to Chapters 5 and 6, returning later to Chapters 3 and 4. It would also be possible to omit reading Chapter 2 completely if the reader is not interested in specific applications of the theory of stochastic processes.

Another possibility is to read Chapters 5 and 6 before Chapter 1 followed by the rest of the book.

Some points should also be made regarding the Bibliography presented at the end of the work.

The Bibliography is intended as an introductory and not as a complete one. It is thus oriented more towards suggesting a "direction for learning" rather than giving emphasis to the authors who have contributed most, at a scientific level, to the development of the different aspects of the subject treated here.

1

Theory of Markovian Stochastic Processes

1.1 ELEMENTS OF STATISTICAL MECHANICS

Before discussing Brownian motion and the theory of stochastic processes we will present a brief outline of statistical mechanics. While not attempting to give a rigorous and exhaustive treatment, it will be more complete if we recall here some of the fundamental concepts.

The problem before us is that of describing macroscopic systems, composed of numerous interacting microscopic subsystems (the number of subsystems is of the order of magnitude of Avogadro's number, i.e. 10^{23}). Specifically, we will consider the case in which the subsystems are atoms or molecules.

Our knowledge of the fundamental laws of the microscopic world, which regulate the interaction between atoms and molecules, is such that we can write equations of evolution of a system. However, the complexity and the number of the resulting equations make a direct attack on the problem completely impracticable. To give only an idea of the enormity of the numbers in play, it can be estimated that only the print-out of the initial positions and velocities of the barycenters of the molecules in one mole of gas, using a high-speed printer, would require a period of time of the order of the estimated age of the universe. The microscopic approach holds, therefore, no hope of success: atomic and molecular theory shows us a universe of incessant motion, where every molecule moves along extremely complicated trajectories.

On the other hand, macroscopic observations show the existence of regular and predictable behaviours along with the possibility of describing macroscopic systems using only a few macroscopic variables (that we will call, in a general way, order parameters). This term is normally used in a more precise manner, indicating a macroscopic variable which, during a phase transition, passes from the value zero to a non-zero value. We will here adopt a wider definition, which will be clarified later.

Let us consider, for example, the case of a perfect gas. It is known that on leaving the system isolated for a sufficiently long period of time, it will tend to present constant and spatially homogeneous characteristics which can be described using only three quantities, pressure P, volume V and temperature T, related by the equation:

$$PV = NkT \tag{1.1}$$

where N is the number of molecules present and k a is a universal constant (Boltzmann constant).

Generally, it is an experimental fact that isolated systems tend towards a constant and homogeneous state of thermodynamic equilibrium, in a way independent of the initial conditions. This is due to the fact that, in reality, using macroscopic observations we follow the course of only a few variables, whereas a microscopic description involves an enormous number of variables: it thus happens that a number of microscopic states correspond to the same macroscopic state.

We can give more precision to these considerations by introducing some geometrical concepts. Here and throughout this text we will make our considerations always within the framework of classical mechanics, even though quantum mechanics would be more suited to the description of objects of atomic dimensions. Classical mechanics holds the advantage of being more simple, and the substance of the arguments, which we intend to illustrate, does not greatly differ in either case.

We can therefore give a microscopic description of our physical system in terms of N generalised coordinates and N conjugate moments. A microstate of the system can therefore be represented by a point in a $2N$-dimensional Euclidean space, where the axes carry the generalised coordinates $q_1, \ldots, q_N$ and the moments $p_1, \ldots, p_N$ (the so-called phase space, or Γ-space). In a way corresponding to given initial conditions, the system will evolve in time describing a trajectory in Γ-space, a projection of which on the plane q_k, p_k is shown in Fig. 1.1.

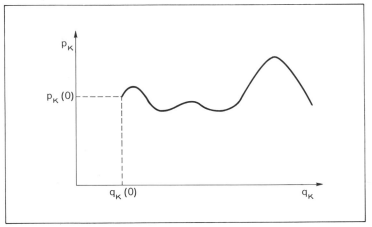

Fig. 1.1 - **Projection onto the plane (q_k, p_k) of the phase space trajectory of a physical system.**

One problem which we have not yet considered is the determination of the initial conditions, which is clearly impossible even in a classical context, because of the number of degrees of freedom of the system. What we do know about the system is limited in practice, to only a few macroscopic constraints, as for example the value of the total energy. What we can not know is the precise point (microstate), in the region of Γ- space individuated by these constraints, in which the system finds itself. We are therefore obliged to adopt a probabilistic point of view, even though this constitutes a clean break with the techniques employed in microscopic physics.

It is now useful to introduce the concept of a Gibbs ensemble. Suppose we have a macroscopic physical system, at time t, and, associated with this, M mental copies of the system, where M is large (it can tend to infinity). Every mental copy satisfies the constraints of the macroscopic system, whereas the different copies find themselves in different microscopic states. The set of M copies is called a Gibbs ensemble and corresponds to a cloud of points in Γ-space.

The idea behind the use of Gibbs ensembles is that since each set represents a different microscopic realisation of the same macroscopic state, it is possible to associate macroscopic quantities to averaging operations on the Gibbs ensemble. Effectively, every macroscopic measurement has a certain duration and constitutes a kind of time average, taken over several successive microstates. Whether it is justifiable or not to substitute ensemble averages with time averages (ergodic hypothesis) is a subtle and much-debated problem into the merits of which we will not enter, noting only that the use of representative ensembles leads to correct results for the equilibrium properties of most physical systems.

We will now consider the specific case of the so-called microcanonical ensemble, which represents an isolated system of constant energy E, known with a level of uncertainty Δ, which reflects the imprecision of the macroscopic measurements. Therefore, if $H(q, p)$ is the Hamiltonian of the system, we have

$$E \leqslant H(q, p) \leqslant E + \Delta \tag{1.2}$$

Since we know nothing of the microstates, the most logical choice is that of attributing the same probability to all the microstates compatible with the condition expressed in Eq. (1.2). In this way we obtain the microcanonical distribution:

$$
\begin{aligned}
\rho(q, p) &= \text{constant} && E \leqslant H(q, p) \leqslant E + \Delta \\
\rho(q, p) &= 0 && \text{elsewhere}
\end{aligned}
\tag{1.3}
$$

where $\rho(q, p)$ represents the density of representative points in Γ-space. The volume occupied by the whole microcanonical ensemble is therefore

$$\Gamma(E) = \int_{\substack{\text{over all} \\ \text{space}}} \rho(q, p)\, dq\, dp \tag{1.4}$$

Calling $\Sigma(E)$ the volume of Γ-space bounded by a surface at constant energy E:

$$\Sigma(E) = \int_{H(q, p) < E} dq\, dp \tag{1.5}$$

and adopting the microcanonical distribution, we find that the volume occupied by the microcanonical ensemble is

$$\Gamma(E) = \Sigma(E + \Delta) - \Sigma(E) \tag{1.6}$$

By putting $\Delta \ll E$, this can be rewritten as:

$$
\left\{
\begin{aligned}
\Gamma(E) &= \omega(E)\Delta \\
\omega(E) &= \frac{\partial}{\partial E} \Sigma(E)
\end{aligned}
\right.
\tag{1.7}
$$

where $\omega(E)$ is the density of states. One can also demonstrate that the function

$$S(E, V) = k \ln \Gamma(E) \tag{1.8}$$

possesses all the properties of entropy. The demonstration is based upon an analysis of the case of two subsystems at first isolated which are then brought together to interact. Such topics can be found in standard textbooks on statistical mechanics.

Let us now consider the case of an isolated system, represented by a certain microcanonical ensemble, composed of two interacting subsystems, one of which is much smaller than the other. In this way we intend to reproduce the case of a system in thermal contact with a heat source. Considering the total microcanonical ensemble, it can be demonstrated that the probability of a given microstate of the smaller system is a function only of its energy. If $H(q, p)$ is the Hamiltonian of the subsystem, the corresponding probability density is

$$\rho(q, p) = \text{constant} \times \exp\left[-H(q, p)/kT\right] \tag{1.9}$$

This is known as the canonical (or Boltzmann) distribution, and can be used to construct directly a Gibbs ensemble for a system in thermal contact with a heat reservoir (canonical ensemble).

Finally, we note that the relationship which links entropy to volume in phase space can be physically re-interpreted. As all states are equally probable in the microcanonical ensemble, the volume of phase space can be considered a measure of the probability of finding the system with a given energy E, obviously where the constraint of constant energy has not been imposed. Thus, the relationship between the entropy and volume in phase space can be re-interpreted as a relationship between the entropy and the probability $p(E)\,dE$ of finding the system with an energy between E and $E + dE$:

$$S(E) = k \ln p(E)$$

Inverting this relationship, we obtain the so-called Einstein equation, which expresses the probability of obtaining a certain macroscopic state as a function of its entropy:

$$p = \exp\left[S/k\right] \tag{1.10}$$

Other macroscopic variables, besides energy, may be used.

In conclusion, we repeat that we have limited ourselves to giving only a brief outline of statistical mechanics without attempting to present a full treatment of the subject and without entering into the subtle and fascinating questions of a methodological and epistemological nature that this subject would merit.

The expressions that will be used throughout the volume are:
—   the Boltzmann distribution (Eq. (1.9))
—   the Einstein equation (Eq. (1.10)).

1.2 BROWNIAN MOTION AND THE LANGEVIN EQUATION

We shall now study the motion of a Brownian particle in a fluid. As usual we will use a classical and not a quantum description. The second law of dynamics gives us

$$M\vec{\dot{v}} = \vec{F}_E(\vec{x}) + \vec{F} \tag{1.11}$$

where M, $\vec{x}$ and $\vec{v}$ indicate the mass, position and velocity, respectively, of the Brownian particle; $\vec{F}_E$ is the external force due to the action of a field (gravitational, electric, etc.) upon the Brownian particle, and which would be present even in the absence of the fluid. The term $\vec{F}$ is the resultant of the forces exerted upon the particle by the fluid molecules:

$$\vec{F} = \sum_i \vec{f}_i \tag{1.12}$$

Naturally, there is no hope of directly resolving such a complex problem and so we must make some simplifying hypotheses.

If the Brownian particle were to be considered as a macroscopic object, hydrodynamics could provide some indication on how to confront the problem. An effective way of representing the interaction between a fluid and a moving macroscopic object is to introduce a frictional force proportional to the velocity

$$\vec{F}_{fr} = -\gamma \vec{v} \tag{1.13}$$

where γ is the friction coefficient. Unfortunately, such a representation is inadequate in the case of Brownian motion.

In fact, if we take the case where the external force is zero, we would obtain an exponentially decreasing velocity, thus excluding the possibility of describing the irregular and incessant motion of the Brownian particle.

Langevin understood that it was possible to arrive at a physically significant description of Brownian motion by adding a fluctuating force to the deterministic friction term. The famous Langevin equation, when the external force is zero, is therefore:

$$\vec{\dot{v}} = (\vec{F}/M) = -\beta\vec{v} + \vec{A}(t) \tag{1.14}$$

The fluctuating force $\vec{A}(t)$ represents the resultant of the molecular collisions, the usual frictional term having been "extracted". Physically, the "friction" derives from the fact that the momentum transferred from the molecules of the fluid to the particle is, on the average, greater in the direction opposing the motion than in the direction of the motion. This is why there is a progressive "slowing-down" of a macroscopic body in a fluid, with the consequent dissipation of energy.

In the case of Brownian motion, we have essentially the same phenomenon (molecule-particle collisions), but the appearance is completely different. This is because the Brownian particle is so small and light that it "feels" the changes in direction, from the average value, of the resultant of the forces, so that it undergoes an erratic motion of the type shown in Fig. 1.2.

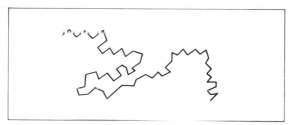

Fig. 1.2 - Two-dimensional trajectory of a Brownian particle.

We will call the mesoscopic level of describing the physical world that which the Brownian particle demonstrates, reserving the term macroscopic for objects, the behaviour of which can be adequately described using deterministic equations.

The Langevin equation (Eq. (1.14)) has a rather particular meaning, which makes it different from ordinary differential equations. In fact, the fluctuating force $\vec{A}(t)$ is not considered as a given function of time, but is defined only in statistical terms. This brings about a corresponding change in the meaning of the solution of Eq. (1.14): that is, to find the probability distribution of the variable in question, and no longer its law of motion.

Let us now try to define the statistical properties of the fluctuating force. From experimental observations and from considerations of symmetry, it must have an average value of zero:

$$\langle \vec{A}(t) \rangle = 0 \tag{1.15}$$

This is because we have already taken into account the only element that could have brought about a disruption of the symmetry, that is the relative particle-fluid motion, by using the dissipative term $-\gamma\vec{v}$. To clarify our ideas, we can consider the averages as ensemble averages, calculated over a population of Brownian particles which are not subject to mutual interaction. Observations also demonstrate that the direction and intensity of a change in the motion of a Brownian particle appear to be independent of the characteristics of the preceding change. Let us suppose therefore that:

$$\langle A(t)A(t+s)\rangle = \phi(t, s) \tag{1.16}$$

where $\phi(t, s)$ is appreciably different from zero only in a small neighbourhood around the value $s = 0$. We have abolished the vectorial notation in order not to complicate matters, and from here onwards we will limit ourselves to the one-dimensional case, without this affecting our fundamental reasoning. The l.h.s. of Eq. (1.16) is the so-called auto-correlation function of the stochastic force $A(t)$. This measures the duration of the "memory" of the fluctuating force. In fact, if the interval, s, is sufficiently long to guarantee the independence of $A(t + s)$ from $A(t)$, the average is broken down into the product of two averages, which is zero because of Eq. (1.15). We will now consider the case in which the memory is infinitely short, so that

$$\langle A(t)A(t+s)\rangle = \sigma^2 \delta(s) \tag{1.17}$$

where $\delta(s)$ is the well-known Dirac's delta distribution. We see that writing the previous equation with

$$\sigma^2 = \langle A^2(t)\rangle \tag{1.18}$$

constant, we have assumed that the mean square value of the fluctuating force does not vary with time.

The hypothesis of a δ-correlated fluctuating force, also expressed by the term "white noise", introduces a "pathological" element into our mathematical description; we should therefore proceed with caution. In particular, one notes that a delta-correlated fluctuating force can not even be drawn. In fact, the auto-correlation function is obtained by summing the value of the quantity $A(t)A(t+s)$ over the particles that make up the statistical ensemble to which the averaging operation refers:

$$\phi(s) = \sum_k A_k(t)A_k(t+s) \tag{1.19}$$

If one were able to draw the temporal behaviour of $A_k(t)$, then, given a sufficiently small value of s, in the vast majority of cases $A_k(t + s)$ would have the same sign as $A_k(t)$; thus, almost all of the contributions to the summation would be positive, and $\phi(s)$ would have non-zero values for small but non-zero values of s, contrary to the limiting hypothesis of Eq.(1.17). The impossibility of describing the temporal behaviour of $A(t)$ is reflected by the analogous impossibility for $\dot{v}(t)$, which turns out to be proportional to the stochastic force. The fact that $\dot{v}$ is not a function, but a "pathological object", implies in its turn that v can not be differentiated at any point.

Clearly, this is an idealization. If we were to observe the physical motion of the Brownian particle over time intervals less than the average time between one collision and the next (typically 10^{-11} seconds), we would, in fact, observe a linear uniform motion. If, however, such small intervals are not of interest, we can hope that the approximate description, in terms of a delta-correlated fluctuating force, will be sufficient for all practical purposes.

In reality, therefore, the limit for $t \to 0$ is taken physically as meaning the consideration of very short, but not infinitesimal, periods of time. The key point is that the dynamic behaviour of $v(t)$ must be much slower than that of the fluctuating force. In this way it is possible to interpret physically the Langevin equation as a convenient notation really meaning that

$$\Delta v(t) = -\beta v(t)\Delta t + \Delta w(t)$$
$$\Delta w(t) = \int_t^{t+\Delta t} A(s)\,ds \tag{1.20}$$

The hypothesis that two separate time scales exist is implicit in this equation, since we have substituted

$$-\int_t^{t+\Delta t} v(s)\,ds$$

with

$$-v(t)\,\Delta t$$

The preceding equation can also be written in the form

$$dv(t) = -\beta v(t)\,dt + dw(t)$$

$$w(t) = \int_0^t A(s)\,ds \tag{1.21}$$

where the symbol "d" is taken as representing a "physical infinitesimal".

The integral of the fluctuating force $A(t)$, that is $w(t)$, is known as a "Wiener process", which is sometimes given the name "Brownian motion", the reason for which we shall see later.

To come to terms with the problem of finding the probability distribution for the velocity and the position of the Brownian particle, we have yet to define the statistical properties of the fluctuating force $A(t)$, or, equivalently, of the Wiener process $w(t)$. So far, we have only determined the first two moments of the distribution (Eqs. (1.15) and (1.16)). Note that these two equations also determine the first two moments of the distribution of $w(t)$. In fact:

$$\langle w(t) \rangle = \int_0^t \langle A(s) \rangle\,ds = 0 \tag{1.22}$$

$$\langle w(t)w(s) \rangle = \int_0^t dx \int_0^s dy \,\langle A(x)A(y) \rangle = \sigma^2 \min(t, s) \tag{1.23}$$

Now, there is a known distribution, the Gaussian distribution, which is completely determined by its first two moments. Besides, the central-limit theorem essentially states that the resultant of infinite independent contributions follows a Gaussian distribution. Given that $w(t)$ is the sum of many uncorrelated collisions, it is therefore reasonable to suppose that it effectively follows a Gaussian distribution, which is therefore determined by Eqs. (1.22) and (1.23). If $p(w)dw$ is the probability of finding a noise level between w and $w + dw$, then,

$$p(w) = \frac{1}{\sqrt{2\pi\sigma^2 t}}\, e^{-w^2/2\sigma^2 t} \tag{1.24}$$

In this way we have completed the definition of the statistical properties of $A(t)$ and of $w(t)$, and we can now turn our attention to the problem of finding the probability distribution of the position x and the velocity v.

1.3 BASIC THEORY OF STOCHASTIC PROCESSES

At this point it is necessary to introduce some fundamental concepts of probability theory. As many books on the subject are available (Papoulis' book, cited in the Bibliography, is particularly clear), here we will present only a brief description. We will limit ourselves to some of the most important definitions and properties of the theory of random variables but will give a wider treatment of the theory of stochastic processes.

It is possible to give a rigorous definition to the key concept of a random experiment, but it will be sufficient for our purposes to refer to the intuitive notion of what a random experiment involves. The results, or better, the sets of results from a random experiment, are called events and it is possible to associate a probability with each event.

A random variable is a function of the results of a random experiment whose range can be the set of real numbers, one of its subsets, or the set of complex numbers, etc. Henceforth we shall limit ourselves to the case of real numbers of which the order relation "greater than" can be established.

We also give, briefly, the definitions of cumulative probability $P(y)$ and of probability density $p(y)$. We can associate a random variable x with a cumulative probability

$$P(y) = Pr\{x \leqslant y\}$$

where $Pr\, A$ is the probability of event A and $\{x \leqslant y\}$ is the set of all experimental results associated with a value not greater than y for the variable x. We can also define a probability density:

$$p(y) = dP(y)/dy$$

Thus $p(y)\,dy$ is the probability of finding the random variable x in the range $[y, y + dy]$. It is to be noted that, if x is a continuous variable, the probability of obtaining a particular value is precisely nil, and it only has sense to ask what the probability is of obtaining values in the interval dy.

A stochastic process is the correspondence between a random experiment and a group of functions. In other words, the result of an experiment is associated with a function, belonging to a certain set. Physically, a stochastic process represents the evolution in time of an irregular function, the course of which at most can be predicted in terms of probability, as for example the velocity of a Brownian particle, the voltage changes across a resistor, etc.; in any case, in the mathematical definition of a stochastic process no reference is made to the irregular behaviour of the intervening functions.

It is now straightforward to introduce the concept of a vectorial stochastic process where each experimental result is a set of values corresponding to the components of a vector. Now we are faced with the problem of generalising the concept of probability and the definition of probability density to the case of stochastic processes. Here also, the probability of obtaining exactly any particular value is precisely zero, if we consider continuous variables. We can, however, formulate the problem in non-trivial terms by considering discrete moments in time $t_1, t_2 \ldots t_n$, and asking ourselves what the probability is of finding $x(t_k)$ between y_k and $y_k + dy_k$. We will indicate the corresponding probability density by

$$p_n\begin{pmatrix} y_n \cdots y_1 \\ t_n \cdots t_1 \end{pmatrix} dy_1 \ldots dy_n = Pr\{y_1 \leqslant x(t_1) \leqslant y_1 + dy_1, \ldots, y_n \leqslant x(t_n) \leqslant y_n + dy_n\} \quad (1.25)$$

The density of Eq. (1.25) is an n-state joint probability density. It is also worthwhile introducing the concept of transition probability density with the following relationship:

$$p_n\begin{pmatrix} y_n \cdots y_1 \\ t_n \cdots t_1 \end{pmatrix} = p\begin{pmatrix} y_n \\ t_n \end{pmatrix}\begin{pmatrix} y_{n-1} \cdots y_1 \\ t_{n-1} \cdots t_1 \end{pmatrix} p_{n-1}\begin{pmatrix} y_{n-1} \cdots y_1 \\ t_{n-1} \cdots t_1 \end{pmatrix} \quad (1.26)$$

That is, the probability of finding the value y_1 at time t_1, y_2 at time t_2, ..., y_n at time t_n, is equal to the probability of finding y_1 at time t_1, ..., y_{n-1} at time t_{n-1} multiplied by the probability of finding y_n at t_n, given the certainty that the system was in y_1 at t_1, ... y_{n-1} at t_{n-1}. Intuitively, we can say that the transition (probability) density "propagates" the probability since, given the joint probability density for $n-1$ states, this allows one to obtain that for

n states. It is clear that the statistical description of a stochastic process requires knowledge of the complete hierarchy of the joint probabilities for n points ($n = 1, 2, 3 \ldots , n \to \infty$):

$$p_n \begin{pmatrix} y_n & y_{n-1} & \cdots & y_1 \\ t_n & t_{n-1} & \cdots & t_1 \end{pmatrix} \tag{1.27}$$

The hierarchy $p_1, p_2 \ldots$ corresponds to an increasingly detailed knowledge of the process. Experimentally, the way to determine this is to record a large number of time histories $x(t)$, relating to systems prepared in the same manner (speaking, naturally, from the macroscopic point of view). In order to find

$$p_1 \begin{pmatrix} y \\ t \end{pmatrix}$$

one determines the relative frequency with which values between y and $y + dy$ occur at time t. In order to find

$$p_2 \begin{pmatrix} y_2 & y_1 \\ t_2 & t_1 \end{pmatrix}$$

one must count the number of times, in the same recording, that values close to y_1 at time t_1 occur and, simultaneously, the number of times that values close to y_2 at time t_2 occur. One proceeds in the same way for succeeding terms along the hierarchy of joint probability densities.

This hierarchy forms the basis for classifying stochastic processes. The simplest case is that of a purely random process, in which the values of the variable at successive moments in time are uncorrelated, so that the joint probability is expressed as the product of the probabilities of single events:

$$p_2 \begin{pmatrix} y_2 & y_1 \\ t_2 & t_1 \end{pmatrix} = p_1 \begin{pmatrix} y_2 \\ t_2 \end{pmatrix} p_1 \begin{pmatrix} y_1 \\ t_1 \end{pmatrix} \tag{1.28}$$

etc. In this case, p_1 obviously contains all the information necessary for defining the process.

The next case, in order of complexity, is that of the so called "Markovian processes", whose definition is particularly simple if expressed in terms of transition densities. A stochastic process is considered Markovian if, $\forall n \geqslant 2$:

$$p \begin{pmatrix} y_n & \vline & y_{n-1} & \cdots & y_1 \\ t_n & \vline & t_{n-1} & \cdots & t_1 \end{pmatrix} = p \begin{pmatrix} y_n & \vline & y_{n-1} \\ t_n & \vline & t_{n-1} \end{pmatrix} \tag{1.29}$$

Intuitively, the system has a short memory: the value of the variable x at time t_{n-1} determines the probability of finding other values at the following time t_n, whereas the preceding history (t_{n-2}, t_{n-3}, etc.) does not have any influence on future statistical properties.

One could even consider more complex cases in which the information is contained in the knowledge of

$$p_3 \begin{pmatrix} y_3 & y_2 & y_1 \\ t_3 & t_2 & t_1 \end{pmatrix}$$

and so on. Anyway, in these cases, as we shall see later in Chapter 3 on non-Markovian processes, it is better to resort to the Markovian case, enlarging the definition of the state of the system by the introduction of further variables. In this way the original non-Markovian process is considered as a kind of projection of a larger Markovian process.

We will now demonstrate an important property: if a process is Markovian, knowing the probability density of a single event

$$p_1\begin{pmatrix} y_1 \\ t_1 \end{pmatrix}$$

and the transition density

$$p\begin{pmatrix} y_2 & y_1 \\ t_2 & t_1 \end{pmatrix}$$

is sufficient to define the whole hierarchy of joint probabilities, and therefore to completely characterise the process in a statistical sense. This is simple enough to demonstrate. By definition:

$$p_2\begin{pmatrix} y_2 & y_1 \\ t_2 & t_1 \end{pmatrix} = p\begin{pmatrix} y_2 & y_1 \\ t_2 & t_1 \end{pmatrix} p_1\begin{pmatrix} y_1 \\ t_1 \end{pmatrix}$$

Thus p_2 can be determined. Then

$$p_3\begin{pmatrix} y_3 & y_2 & y_1 \\ t_3 & t_2 & t_1 \end{pmatrix} = p\begin{pmatrix} y_3 & y_2 \\ t_3 & t_2 \end{pmatrix} p_2\begin{pmatrix} y_2 & y_1 \\ t_2 & t_1 \end{pmatrix} = p\begin{pmatrix} y_3 & y_2 \\ t_3 & t_2 \end{pmatrix} p\begin{pmatrix} y_2 & y_1 \\ t_2 & t_1 \end{pmatrix} p_1\begin{pmatrix} y_1 \\ t_1 \end{pmatrix}$$

and for the *n*-time density we will obtain:

$$p_n\begin{pmatrix} y_n & \cdots & y_1 \\ t_n & \cdots & t_1 \end{pmatrix} = p\begin{pmatrix} y_n & y_{n-1} \\ t_n & t_{n-1} \end{pmatrix} p\begin{pmatrix} y_{n-1} & y_{n-2} \\ t_{n-1} & t_{n-2} \end{pmatrix} \cdots p\begin{pmatrix} y_2 & y_1 \\ t_2 & t_1 \end{pmatrix} p_1\begin{pmatrix} y_1 \\ t_1 \end{pmatrix}$$

Here we add another important definition: a stochastic process is stationary if the probability density of a single event is not time-dependent and if the joint densities depend only upon the time difference between observations:

$$\frac{\partial}{\partial t} p_1\begin{pmatrix} y \\ t \end{pmatrix} = 0 \tag{1.30a}$$

$$p_n\begin{pmatrix} y_n & \cdots & y_1 \\ t_n & \cdots & t_1 \end{pmatrix} = p_n\begin{pmatrix} y_n & \cdots & y_1 \\ t_n + \Delta & \cdots & t_1 + \Delta \end{pmatrix} \tag{1.30b}$$

In the case of a Markovian process, the latter condition can be summarised by the following simple expression:

$$p\begin{pmatrix} y_2 & y_1 \\ t_2 & t_1 \end{pmatrix} = p\begin{pmatrix} y_2 & y_1 \\ t_2 + \Delta & t_1 + \Delta \end{pmatrix} \tag{1.31}$$

In the case where Eq. (1.31) or (1.30b) are satisfied, without conditions being imposed on

$$p_1\begin{pmatrix} y \\ t \end{pmatrix}$$

the process is said to be time homogeneous. Time homogeneity expresses the invariance of the mechanism which generates fluctuations. It is to be noted that in the case of stationary stochastic processes it is true to say that time averages and ensemble averages are coincident.

Let us now try to decide whether Brownian motion is or is not a Markovian process. We must first clarify a terminological ambiguity: the term "Brownian motion" can, in fact, be used to refer to a number of stochastic processes. If we take the position of a Brownian particle immersed in a fluid, then we are considering the process $x(t)$. Whereas, if we turn our attention to the velocity of the particle, we are considering the process $v(t)$. We can also take into consideration the multidimensional process $[x(t), v(t)]$. In the mathematical literature, however, it is the Wiener process $w(t)$ which is identified as Brownian motion. Let us now examine the Markov property of these different processes.

As far as the Wiener process is concerned, from the definition

$$w(t) = \int_0^t A(s)\,ds$$

we obtain:

$$w(t_n) = \int_0^{t_n} A(s)\,ds = w(t_{n-1}) + \int_{t_{n-1}}^{t_n} A(s)\,ds \tag{1.32}$$

Since $A(s)$ is delta-correlated, its values in the interval $[t_{n-1}, t_n]$ are independent of the previous history, and $w(t_n)$ does not depend upon the history preceding t_{n-1}. Therefore the Wiener process is Markovian. In fact, (1.32) demonstrates a stronger property than its Markovian character, i.e. that the process is one of independent increments. This means that the statistical distribution of the increments $\Delta w(t) = w(t + \Delta t) - w(t)$ is independent of the value of $w(t)$, besides being independent of all previous values.

Turning now to the stochastic process $v(t)$, we see that the formal solution to the Langevin equation (1.14), with the initial condition $v(0) = 0$, is:

$$v(t) = e^{-\beta t} \int_0^t e^{\beta s} A(s)\,ds$$

from which

$$v(t + \Delta t) = e^{-\beta t} e^{-\beta \Delta t} \left[\int_0^t e^{\beta s} A(s)\,ds + \int_t^{t+\Delta t} e^{\beta s} A(s)\,ds \right] =$$

$$= e^{-\beta \Delta t} v(t) + e^{-\beta(t + \Delta t)} \int_t^{t+\Delta t} e^{\beta s} A(s)\,ds$$

One can then see that the knowledge of $v(t)$ and of the fluctuating force between t and $t + \Delta t$ determines $v(t + \Delta t)$, and therefore $v(t)$ is a Markovian process. Also, the increment

$$\Delta v(t) = v(t + \Delta t) - v(t) = (e^{-\beta \Delta t} - 1) v(t) + e^{-\beta(t + \Delta t)} \int_t^{t+\Delta t} e^{\beta s} A(s)\,ds$$

depends upon $v(t)$, and therefore we are not dealing with a process involving independent increments.

Finally, let us consider the stochastic process $x(t)$:

$$x(t) = \int_0^t v(s)\,ds$$

$$x(t + \Delta t) = \int_0^{t+\Delta t} v(s)\,ds = x(t) + \int_t^{t+\Delta t} v(s)\,ds$$

One can thus see that $x(t)$, taken by itself, is not a Markovian process. In fact, in order to find $x(t + \Delta t)$ we need to know the values of velocity in the interval $[t, t + \Delta t]$. From the preceding relationship one can immediately deduce that the vectorial process $[x, v]$ is Markovian.

It is to be noted that this is the first example we have met of a non-Markovian stochastic process, $x(t)$, that "becomes Markovian on adding variables", or that can be considered as a projection of a process $[x, v]$ in the subspace of only the x variable.

To conclude, we now determine the relationship that exists between the joint probability density for n states and that for $n + 1$ states. Let us consider the following two quantities:

$$p_n \begin{pmatrix} y_n & \cdots & y_k & y_{k-1} & \cdots & y_1 \\ t_n & \cdots & t_k & t_{k-1} & \cdots & t_1 \end{pmatrix}, \quad \text{and} \quad p_{n+1} \begin{pmatrix} y_n & \cdots & y_k & y_L & y_{k-1} & \cdots & y_1 \\ t_n & \cdots & t_k & t_L & t_{k-1} & \cdots & t_1 \end{pmatrix}$$

One can see that the extrema are the same in both cases. The $(n + 1)-$state probability contains more information than the n-state probability as it corresponds to a more detailed observation of the system, having one extra "gate". It is clear, therefore, that one obtains the n-state density by integrating over all possible values of the extra observed state:

$$p_n \begin{pmatrix} y_n & \cdots & y_k & y_{k-1} & \cdots & y_1 \\ t_n & \cdots & t_k & t_{k-1} & \cdots & t_1 \end{pmatrix} = \int p_{n+1} \begin{pmatrix} y_n & \cdots & y_k & y_L & y_{k-1} & \cdots & y_1 \\ t_n & \cdots & t_k & t_L & t_{k-1} & \cdots & t_1 \end{pmatrix} dy_L \qquad (1.33)$$

1.4 THE CHAPMAN-KOLMOGOROV AND FOKKER-PLANCK EQUATIONS

In this Section we will show how it is possible to associate Markovian stochastic processes with suitable partial differential equations, that describe the temporal evolution of the probability density. In this way the "pathological" Langevin equation is substituted by more customary and easily managed expressions of mathematical physics.

Our point of departure will be the last equation of the preceding Section (Eq. (1.33)), that links the n-state joint density p_n to p_{n+1}. When the process is Markovian, taking $n = 2$ we obtain

$$p_2 \begin{pmatrix} y_3 & y_1 \\ t_3 & t_1 \end{pmatrix} = \int p_3 \begin{pmatrix} y_3 & y_2 & y_1 \\ t_3 & t_2 & t_1 \end{pmatrix} dy_2$$

with

$$p_2 \begin{pmatrix} y_3 & y_1 \\ t_3 & t_1 \end{pmatrix} = p \begin{pmatrix} y_3 \\ t_3 \end{pmatrix} \begin{vmatrix} y_1 \\ t_1 \end{vmatrix} p_1 \begin{pmatrix} y_1 \\ t_1 \end{pmatrix}$$

$$p_3 \begin{pmatrix} y_3 & y_2 & y_1 \\ t_3 & t_2 & t_1 \end{pmatrix} = p \begin{pmatrix} y_3 \\ t_3 \end{pmatrix} \begin{vmatrix} y_2 \\ t_2 \end{vmatrix} p \begin{pmatrix} y_2 \\ t_2 \end{pmatrix} \begin{vmatrix} y_1 \\ t_1 \end{vmatrix} p_1 \begin{pmatrix} y_1 \\ t_1 \end{pmatrix}$$

From this it follows that

$$
p\left(\begin{matrix} y_3 \\ t_3 \end{matrix} \middle| \begin{matrix} y_1 \\ t_1 \end{matrix}\right) = \int p\left(\begin{matrix} y_3 \\ t_3 \end{matrix} \middle| \begin{matrix} y_2 \\ t_2 \end{matrix}\right) p\left(\begin{matrix} y_2 \\ t_2 \end{matrix} \middle| \begin{matrix} y_1 \\ t_1 \end{matrix}\right) dy_2
\tag{1.34}
$$

The equation that we have now obtained is known as the Chapman-Kolmogorov equation. This expresses the fact that the transition probability density from state 1 to state 3 is the sum (integral), over all possible intermediate states, of the probability of transition $1 \to 2 \to 3$.

Note that Eq. (1.34) is a closed equation, in the sense that the only variables which appear are the two-state transition densities.

One can obtain an equation of an analogous form for the probability density of a single event, losing however this "closure" property. Following the same reasoning as above for the case $n = 1$, we obtain

$$
p_1\left(\begin{matrix} y \\ t \end{matrix}\right) = \int p\left(\begin{matrix} y \\ t \end{matrix} \middle| \begin{matrix} y' \\ t' \end{matrix}\right) p_1\left(\begin{matrix} y' \\ t' \end{matrix}\right) dy'
\tag{1.35}
$$

We can now deduce, from the Chapman-Kolmogorov equation, an extremely important partial differential equation for the transition density, which is associated with the names of Fokker and Planck. Let us consider a time-homogeneous Markovian process, that is, for which the transition densities depend only upon the difference between the two instants of observation and, in order to simplify the notation, let us put

$$
p(y, t \mid y_0) = p\left(\begin{matrix} y \\ t \end{matrix} \middle| \begin{matrix} y_0 \\ 0 \end{matrix}\right) = p\left(\begin{matrix} y \\ t+s \end{matrix} \middle| \begin{matrix} y_0 \\ s \end{matrix}\right)
\tag{1.36}
$$

The Chapman-Kolmogorov equation for a homogeneous process then becomes

$$
p(x, t + \Delta t \mid x_0) = \int p(x, \Delta t \mid y) \, p(y, t \mid x_0) \, dy
\tag{1.37}
$$

We now take an arbitrary function $g(x)$, having all the required mathematical properties (e.g. differentiable to the required order, asymptotic properties, etc.). We therefore multiply both sides of the preceding equation by $g(x)/\Delta t$ and integrate w.r.t. x:

$$
\frac{1}{\Delta t} \int p(x, t + \Delta t \mid x_0) \, g(x) \, dx =
$$
$$
\frac{1}{\Delta t} \iint p(x, \Delta t \mid y) \, p(y, t \mid x_0) \, g(x) \, dx \, dy
\tag{1.38}
$$

We then develop $g(x)$ in Taylor series, about the value $x = y$, to the second order:

$$
g(x) = g(y) + g'(y)(x - y) + (1/2) g''(y)(x - y)^2
$$

Substituting in the r.h.s. of Eq. (1.38) and assuming that the transition probability density is normalised, such that

$$
\int p(x, \Delta t \mid y) \, dx = 1
$$

we have

$$\frac{1}{\Delta t} \iint p(x, \Delta t \mid y)\, p\,(y,\ t \mid x_0)\, g\,(x)\, dx\, dy = \frac{1}{\Delta t} \int dy\ g(y)\, p(y,\ t \mid x_0) +$$

$$+ \frac{1}{\Delta t} \int dy\ p(y,\ t \mid x_0)\, g'(y) \int dx\,(x - y)\, p\,(x,\ \Delta t \mid y) +$$

$$+ \frac{1}{2\Delta t} \int dy\ p\,(y,\ t \mid x_0)\, g''(y) \int dx\,(x - y)^2 p(x,\ \Delta t \mid y)$$

Let us now assume that the two integrals w.r.t. x that appear in this expression exist and take the form:

$$\int (x - y)\, p\,(x,\ \Delta t \mid y)\, dx = A\,(y)\, \Delta t + o(\Delta t)$$

$$\int (x - y)^2\, p\,(x,\ \Delta t \mid y)\, dx = B\,(y)\, \Delta t + o(\Delta t)$$

(1.39)

with

$$\lim_{\Delta t \to 0} o\,(\Delta t)/\Delta t = 0$$

The integrals which appear in Eq. (1.39) are the first two infinitesimal transition moments of our probability distribution. As we will then go to the limit $\Delta t \to 0$, contributions of order higher than Δt will not be of interest.

    Now let us assume that moments whose order is higher than the second are vanishingly small with respect to Δt, so that truncating the development to the second order is justified. This hypothesis, together with that expressed in Eq. (1.39), appears at this point to be completely arbitrary. In reality, as we shall later see, they are completely justified in the case of a system which obeys a Langevin equation. For the moment we will postpone the justification for these adopted hypotheses, as such a discussion here would interrupt the derivation of the Fokker-Planck equation.

    From the equations above, it follows that the Chapman-Kolmogorov equation can be written as:

$$\int dx\ g(x)\ \frac{p\,(x,\ t + \Delta t \mid x_0) - p(x,\ t \mid x_0)}{\Delta t} = \int dy \left[g'(y)\, A(y) + \frac{1}{2}\, g''(y)\, B(y) \right] p(y,\ t \mid x_0)$$

Approaching the limit $\Delta t \to 0$ and re-naming the integration variable of the r.h.s., we obtain:

$$\int dx\ g(x)\ \frac{\partial}{\partial t}\, p(x,\ t \mid x_0) = \int dx \left[g'(x)\, A(x) + \frac{1}{2}\, g''(x)\, B(x) \right] p(x,\ t \mid x_0)$$

Upon partial integration, assuming that the arbitrary function $g(x)$ disappears at infinity along with its first derivative, the r.h.s. then becomes:

$$\int dx\, g(x) \left\{ -\frac{\partial}{\partial x}\, [A\,(x)\, p(x,\ t \mid x_0)] + \frac{1}{2}\, \frac{\partial^2}{\partial x^2}\, [B\,(x)\, p\,(x,\ t \mid x_0)] \right\}$$

Since $g(x)$ is arbitrary, we obtain the Fokker-Planck equation:

$$\frac{\partial}{\partial t}\, p\,(x,\ t \mid x_0) + \frac{\partial}{\partial x}\, [A\,(x)\, p\,(x,\ t \mid x_0)] = \frac{1}{2}\, \frac{\partial^2}{\partial x^2}\, [B\,(x)\, p\,(x,\ t \mid x_0)]$$

(1.40)

As a boundary condition, it is generally required that the probability of finding the system at infinity is zero for every finite value of t:

$$\lim_{|x| \to \infty} p(x, t \mid x_0) = 0 \tag{1.41}$$

Besides, as we are dealing with a transition density, at the limit $t \to 0$ we should obtain the initial value x_0 with probability one, that is, with certainty. Thus

$$\lim_{t \to 0} p(x, t \mid x_0) = \delta(x - x_0) \tag{1.42}$$

Another condition which one can usually impose upon the physically significant solutions of the Fokker-Planck equation is that, at the limit $t \to \infty$, the transition density must tend towards the equilibrium distribution, completely "forgetting" the initial condition. As we shall see in what follows, from this condition we can derive important fluctuation-dissipation theorems.

For the sake of completeness, we note that, if our point of departure had been the equation for the probability of a single event $p(x, t)$ (Eq. (1.35)), we would have obtained an equation of the same form as Eq. (1.40) with $p(x, t)$ in place of the transition density $p(x, t \mid x_0)$. The moments $A(x)$ and $B(x)$ are also defined here by Eqs. (1.39), that is, in terms of the transition densities.

We now give the Fokker-Planck equation for a vectorial stochastic process $z(t)$, without repeating the considerations which are identical to those made for the one-dimensional case:

$$\frac{\partial}{\partial t} p(z, t \mid z_0) + \sum_{i=1}^{N} \frac{\partial}{\partial z_i} \{A_i(z) p(z, t \mid z_0)\} =$$

$$= \frac{1}{2} \sum_{i,j=1}^{N} \frac{\partial^2}{\partial z_i \partial z_j} \{B_{ij}(z) p(z, t \mid z_0)\} \tag{1.43}$$

with:

$$A_i(z) = \lim_{\Delta t \to 0} \frac{1}{\Delta t} \int (z_i' - z_i) p(z', \Delta t \mid z_0) dz_1' \ldots dz_N' \tag{1.44a}$$

$$B_{ij}(z) = \lim_{\Delta t \to 0} \frac{1}{\Delta t} \int (z_i' - z_i)(z_j' - z_j) p(z', \Delta t \mid z_0) dz_1' \ldots dz_N' \tag{1.44b}$$

In this case we have also assumed that the first two transition moments are of the order of Δt and that higher-order moments are zero. We will now show that these assumptions are coherent with the "Langevin approach"; in this way we will obtain correct expressions for the infinitesimal transition moments of the different stochastic processes associated with Brownian motion and we will then derive the corresponding Fokker-Planck equations.

The complete Langevin equations are:

$$\Delta x(t) = v(t) \Delta t$$
$$\Delta v(t) = [-\beta v(t) + \tilde{F}_E(x)] \Delta t + \Delta w(t) \tag{1.45}$$

with $\tilde{F}_E(x) = F_E(x)/m$. We will now consider the case of a constrained Brownian motion, i.e. in the presence of an external force field, since this does not complicate our treatment and allows us to obtain more general results.

Let us first consider the stochastic process $v(t)$. From the definition it follows that:

$$A(v) = \lim_{\Delta t \to 0} \langle \Delta v \rangle / \Delta t$$

$$B(v) = \lim_{\Delta t \to 0} \langle \Delta v^2 \rangle / \Delta t$$

From the second of Eqs. (1.45) we then obtain:

$$A(v) = -\beta v + \tilde{F}_E(x)$$

$$B(v) = \lim_{\Delta t \to 0} \langle \Delta w^2(t) \rangle / \Delta t$$

We also see that from Eqs. (1.17) and (1.20):

$$\langle \Delta w^2(t) \rangle = \int_t^{t+\Delta t} ds \int_t^{t+\Delta t} dq \, \langle A(s) A(q) \rangle = \sigma^2 \Delta t \qquad (1.46)$$

Therefore $B(v) = \sigma^2$ and the Fokker-Planck equation for the process $v(t)$ is

$$\frac{\partial}{\partial t} p(v, t \mid v_0) + \frac{\partial}{\partial v} [(\tilde{F}_E(x) - \beta v) p(v, t \mid v_0)] = \frac{1}{2} \sigma^2 \frac{\partial^2}{\partial v^2} p(v, t \mid v_0) \qquad (1.47)$$

We can also write the equation for the transition density of the two-dimensional process $[x(t), v(t)]$; it is, in fact, sufficient to apply Eqs. (1.43) and (1.44) together with the transition moments obtained from the Langevin equation:

$$A_1 = \lim_{\Delta t \to 0} \langle \Delta x \rangle / \Delta t = v$$

$$A_2 = \lim_{\Delta t \to 0} \langle \Delta v \rangle / \Delta t = \tilde{F}_E(x) - \beta v$$

$$B_{11} = \lim_{\Delta t \to 0} \langle \Delta x^2 \rangle / \Delta t = 0$$

$$B_{12} = \lim_{\Delta t \to 0} \langle \Delta x \, \Delta v \rangle / \Delta t = 0$$

$$B_{21} = B_{12} = 0$$

$$B_{22} = \lim_{\Delta t \to 0} \langle \Delta v^2 \rangle / \Delta t = \langle \Delta w^2 \rangle / \Delta t = \sigma^2$$

Thus the corresponding Fokker-Planck equation is:

$$\frac{\partial}{\partial t} p(x, v, t \mid x_0, v_0) + \frac{\partial}{\partial x} [v p(x, v, t \mid x_0, v_0)] +$$

$$+ \frac{\partial}{\partial v} [(\tilde{F}_E(x) - \beta v) \; p(x, v, t \mid x_0, v_0)] = \frac{1}{2} \frac{\partial^2}{\partial v^2} [\sigma^2 p(x, v, t \mid x_0, v_0)] \qquad (1.48)$$

Finally let us consider another equation with which it is possible to associate Brownian motion using an approximate treatment. We will examine in greater detail considerations of this type further on, whereas here we will limit ourselves to an intuitive treatment. Let us suppose that the fluid with which we are dealing is very viscous, and that the dissipative effects are, therefore, relevant. Formally, this means that β is a very large parameter so that the velocity is a heavily damped variable. Under this hypothesis we are allowed to assume that any variation of it takes place on a much shorter time scale than that of x. Thus putting $\dot{v} \approx 0$ in the deterministic equations

$$\dot{x} = v$$

$$\dot{v} = -\beta v + \tilde{F}_E(x)$$

we get

$$\dot{x} = v$$

$$v = \tilde{F}_E(x)/\beta$$

One then sees that the "fast" variable v, which rapidly relaxes, instantly follows, to this first approximation, the slow variable x. A widespread and appealing terminology calls v the "slave variable" of the "order parameter" x, which regulates the overall behaviour of the system. Eliminating v in the preceding equations results in

$$\dot{x} = \tilde{F}_E(x)/\beta$$

This may be considered as the point of departure for writing a new Langevin equation for only the x variable in the overdamped case:

$$\Delta x(t) = \frac{\tilde{F}_E(x)}{\beta} \Delta t + \Delta \tilde{w}(t) \tag{1.49}$$

The calculation of the transition moments is identical to that for the process $v(t)$, so we will not repeat it here. The Fokker-Planck equation is:

$$\frac{\partial}{\partial t} p(x, t \mid x_0) + \frac{\partial}{\partial x} \left\{ \frac{\tilde{F}_E(x)}{\beta} p(x, t \mid x_0) \right\} = \frac{1}{2} \tilde{\sigma}^2 \frac{\partial^2}{\partial x^2} p(x, t \mid x_0) \tag{1.50}$$

with

$$\tilde{\sigma}^2 = \frac{\langle \Delta \tilde{w}^2 \rangle}{\Delta t} .$$

1.5 SOLUTION OF THE EQUATIONS FOR FREE BROWNIAN MOTION

Let us briefly summarise the results obtained in the previous paragraph, for the specific case of free Brownian motion:

$$dx(t) = v(t)\, dt \tag{1.51a}$$

$$dv(t) = -\beta v(t)\, dt + dw(t) \tag{1.51b}$$

The equation for the evolution of the transition density restricted to velocity space is

$$\frac{\partial}{\partial t} p(v, t \mid v_0) + \frac{\partial}{\partial v} (-\beta vp) = \frac{\sigma^2}{2} \frac{\partial^2}{\partial v^2} p \tag{1.52}$$

The corresponding equation in phase space $[x, v]$ is:

$$\frac{\partial}{\partial t} p(x, v, t \mid x_0, v_0) + \frac{\partial}{\partial x} (vp) + \frac{\partial}{\partial v} (-\beta vp) = \frac{\sigma^2}{2} \frac{\partial^2}{\partial v^2} p \tag{1.53}$$

In the overdamped case, where

$$dx(t) = d\tilde{w}(t) \tag{1.54}$$

we have the so-called Smoluchowski equation (in configuration space):

$$\frac{\partial}{\partial t} p(x, t \mid x_0) = \frac{\tilde{\sigma}^2}{2} \frac{\partial^2}{\partial x^2} p \tag{1.55}$$

We note in passing that Eq.(1.54) explains the reason for which the Wiener process is also called Brownian motion.

Let us first examine the solution in velocity space. Besides imposing the conditions in Eqs. (1.41) and (1.42), that is to disappear at infinity and to be reduced to a delta for $t \to 0$, we also assume that the transition density $p(v, t \mid v_0)$ tends towards the Maxwellian equilibrium distribution for $t \to \infty$ independently of the value of v_0:

$$\lim_{t \to +\infty} p(v, t \mid v_0) = \sqrt{\frac{m}{2\pi kT}} \, e^{-mv^2/2kT} \tag{1.56}$$

The solution meeting all the above requirements turns out to be:

$$p(v, t \mid v_0) = \sqrt{\frac{\beta}{2\pi D(1 - e^{-2\beta t})}} \, e^{-\frac{\beta(v - v_0 e^{-\beta t})^2}{2D(1 - e^{-2\beta t})}} \tag{1.57}$$

where we have introduced the velocity space diffusion coefficient, D, defined as:

$$D = \sigma^2/2 \tag{1.58}$$

The process $v(t)$ defined by the Langevin equation (Eq.(1.51b)) is frequently called the Ornstein-Uhlenbeck process.

The transition density given in Eq. (1.57) has a Gaussian form with an average value equal to

$$\langle v(t) \rangle = v_0 \exp[-\beta t]$$

and with variance

$$\langle (v(t) - \langle v(t) \rangle)^2 \rangle = \frac{D}{\beta} (1 - \exp[-2\beta t])$$

Physically, this means that a numerous population of Brownian particles, all having an initial velocity v_0, tend to evolve in such a way that the average velocity approaches exponentially zero, while the variance increases and tends asymptotically towards the value D/β.

The distribution is modified by random collisions between the particles and the fluid molecules; we can see that the asymptotic value of the variance is largely determined by the ratio between the fluctuating force intensity D and the intensity of the dissipative effects β.

The close relationship between fluctuating forces and dissipative effects is confirmed by the following considerations. By imposing the condition that the Ornstein-Uhlenbeck distribution asymptotically tends towards the Maxwellian equilibrium distribution (Eq. (1.56)), we obtain:

$$D = \beta kT/m \tag{1.59}$$

Therefore, knowing the fluctuating force intensity D is sufficient for the determination of the intensity of the frictional forces β, and also vice versa, as long as the temperature is known. This is the first example we have met of a fluctuation-dissipation theorem. This is very important, also from a practical point of view, since it allows us to reduce the number of independent parameters in our models.

The fact that fluctuation and dissipation are so intimately related is not surprising, since we are dealing with two different aspects of the same phenomenon, that is the random interaction between the Brownian particle and the fluid molecules. The existence of two distinctly separated time scales, one of which defines the rate of variation of the stochastic forces, and the other the particle dynamics, has allowed us to adopt a convenient scheme which separates the usual frictional term, $- \beta v$, from the "white" force $A(t)$. In more complex cases we shall see that it is still possible to separate deterministic and stochastic terms, giving us other forms of relationship between fluctuation and dissipation (see the Chapter on non-Markovian systems).

One could also demonstrate that Eq. (1.57) is a fundamental solution of the Fokker-Planck equation (Eq. (1.52)), that is, tends towards $\delta(v - v_0)$ when $t \to 0$. In the case of free Brownian motion it is also possible to find the fundamental solution of the Fokker-Planck equation in phase space (Eq. (1.53)). For further details the reader is referred to the textbooks cited in the Bibliography. Here we will examine Eq. (1.55), regarding only configuration space, which is valid for the overdamped case. This corresponds to the famous diffusion equation, and its fundamental solution is

$$p(x, t \mid x_0) = \frac{1}{\sqrt{4\pi \tilde{D} t}} e^{- \frac{(x - x_0)^2}{4 \tilde{D} t}} \quad \text{with} \quad \tilde{D} = \frac{\tilde{\sigma}^2}{2} \tag{1.60}$$

This describes a Gaussian probability distribution which broadens with time, with a variance proportional to t. This distribution allows us to elucidate the relationship between Brownian motion and the problem of random walk, which, in its simplest (one-dimensional) form, is the motion of a point forced to shift to the left or to the right by a fixed amount l at instants in time each separated by an interval ϵ. The probability of "stepping" in one direction or the opposite direction is the same. Studying this question one obtains an expression for the transition probability density for this process that, simultaneously making l tend to zero and the frequency of the steps to infinity, exactly reproduces Eq. (1.60). Numerous presentations of the problem of Brownian motion start precisely with this simple example, whereas we have preferred a more physical formulation which begins with considerations on the second law of dynamics applied to the Brownian particle showing then under which physical conditions it leads to the distribution of Eq. (1.60).

Let us now examine a simple and elegant technique for finding the transition density of the Brownian particle without needing to resolve the Fokker-Planck equation, by operating directly upon the Langevin equation:

$$\dot{v} = - \beta v + A(t) \tag{1.61}$$

The use of this technique (which takes the name of Ornstein-Uhlenbeck) is, however, limited to the case with the absence of external forces, which we will illustrate here, or to few other cases, and can not therefore be considered as a substitute for the Fokker-Planck method. Formally integrating Eq. (1.61), we obtain:

$$v(t) = v_0 e^{-\beta t} + e^{-\beta t} \int_0^t e^{\beta s} A(s) \, ds \tag{1.62}$$

From this expression we can directly obtain the first moment of the distribution:

$$\langle v(t) \rangle = v_0 \exp(-\beta t) \tag{1.63}$$

Since we are considering a known initial state, it would be more correct to indicate average values using the symbol $\langle \ \rangle_{v_0}$, which we will exclude here in order to simplify the notation. Squaring both sides of Eq. (1.62) to calculate the second moment, we obtain:

$$v^2(t) = v_0^2 e^{-2\beta t} + e^{-2\beta t} \int_0^t ds \int_0^t dq \, e^{\beta(s+q)} A(s) A(q) + 2 v_0 e^{-2\beta t} \int_0^t e^{\beta s} A(s) \, ds$$

Averaging, and remembering that $\langle A(s) A(q) \rangle = 2D\delta (s-q)$, we have:

$$\langle v^2(t) \rangle = v_0^2 \, e^{-2\beta t} + \frac{D}{\beta} \, (1 - e^{-2\beta t}) \qquad (1.64)$$

It is clear that, knowing the statistical properties of the fluctuating force, it would be possible to determine moments of a higher order of the distribution of v. This is not, however, necessary: from the form of Eq. (1.62), $v(t)$ can be seen to be a linear combination of Gaussian terms added to a deterministic part. Thus v also has a Gaussian distribution, and a knowledge of the first two moments is sufficient for its complete determination. In this way we again obtain the Ornstein-Uhlenbeck distribution (Eq. (1.57)).

It is also possible to obtain the probability distribution for the variable x. In fact, from Eq. (1.62), it follows that:

$$x(t) = x_0 + \int_0^t v(s) \, ds = x_0 + \frac{v_0}{\beta} (1 - e^{-\beta t}) + \int_0^t ds \, e^{-\beta s} \int_0^s dq \, e^{\beta q} A(q) \qquad (1.65)$$

from which

$$\langle x(t) \rangle = x_0 + \frac{v_0}{\beta} \, (1 - e^{-\beta t}) \qquad (1.66)$$

In order to obtain the variance $\langle (x - \langle x \rangle)^2 \rangle$, we can rewrite Eq. (1.65), partially integrating and taking into account Eq. (1.66) for the average value; we thus obtain:

$$x(t) = \langle x(t) \rangle - \frac{e^{-\beta t}}{\beta} \int_0^t e^{\beta s} A(s) \, ds + \frac{1}{\beta} \int_0^t A(s) \, ds$$

Squaring and averaging:

$$\langle x^2(t) \rangle - \langle x(t) \rangle^2 = \frac{D}{\beta^3} (-3 + 2\beta t - e^{-2\beta t} + 4e^{-\beta t}) \qquad (1.67)$$

As x is a linear combination of Gaussian terms, it is also a Gaussian variable, and a knowledge of the first two moments is sufficient for the determination of its distribution:

$$p(x, t \mid x_0, v_0) = \sqrt{\frac{\beta^3}{2\pi D(2\beta t - 3 + 4e^{-\beta t} - e^{-2\beta t})}} \cdot$$

$$\cdot \exp\left[-\frac{\beta^3}{2D} \frac{\{x - x_0 - (v_0/\beta)(1 - e^{-\beta t})\}^2}{2\beta t - 3 + 4e^{-\beta t} - e^{-2\beta t}} \right] \qquad (1.68)$$

This equation gives us the transition density in configuration space, without having introduced the condition of high damping. We will now consider that particular case: let us suppose that β is "large", and that we are dealing with a time scale much longer than the decay times of the variations in the velocity: $t \gg 1/\beta$. We can then, to a first approximation, ignore the exponential terms $\exp(-\beta t)$, $\exp(-2\beta t)$. We can also ignore the term which is constant with respect to the term linear in t, that is also "large" since it is proportional to β. Therefore, we obtain the approximate equation:

$$p(x, t \mid x_0, v_0) \cong \sqrt{\frac{\beta^2}{4\pi Dt}} \; \exp \left[- \frac{\beta^2 (x - x_0 - v_0/\beta)^2}{4Dt} \right] \tag{1.69}$$

Comparing this expression with that obtained previously for the overdamped case (Eq. (1.60)), we see that the initial velocity v_0 was not determined as it is here. We must therefore average our expression over the initial velocities. Since we have made β very large, we can put $v_0/\beta \sim 0$, and we thus, once again, obtain Eq. (1.60), as long as:

$$\tilde{D} = D/\beta^2 = kT/m\beta \tag{1.70}$$

This is the fluctuation-dissipation theorem in configuration space. The last step took into account the analogous theorem in phase space, expressed in Eq. (1.59).

Finally, let us examine the autocorrelation function of the Brownian variable v (it will be recalled that the autocorrelation function of the fluctuating force is a Dirac delta function, Eq. (1.17)). From its definition, and using Eq. (1.62) as well as the fact that the average value of the fluctuating force is zero, we have

$$\phi_v(t) = \langle v(t)\, v(o) \rangle = v^2(o)\, e^{-\beta t}$$

It can thus be seen that the "memory" of the stochastic process $v(t)$ falls exponentially with a time constant of $1/\beta$. Although the fluctuating force instantly forgets its own value, this is not the case for the velocity: the physical origin of the memory mechanism which is at work is naturally the inertia of the Brownian particle.

It is also worth considering the power spectral density, i.e. the Fourier transform of the autocorrelation function. Since the latter is an even function, it is only necessary to consider its cosine transform:

$$\phi_v(t) = \int_0^{+\infty} df\, G(f) \cos(2\pi ft)$$

$$G(f) = \int_0^{+\infty} 4\, dt\, \phi_v(t) \cos(2\pi ft)$$

(The normalisation chosen here is that of Wang and Uhlenbeck; clearly it would be possible to adopt others without changes in the physical considerations). The transform of the autocorrelation function of the fluctuating force is a constant function over the whole spectrum, with a

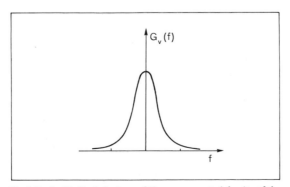

Fig. 1.3 - Qualitative behaviour of the power spectral density of the velocity of a Brownian particle.

value of $4D$: and thus we also have contributions from "infinite frequencies". On the other hand, the power spectrum of the velocity of the Brownian particle takes a Lorentzian form (Fig. 1.3), given by the expression:

$$G_v(f) = 4D/[\beta^2 + (2\pi f)^2]$$

1.6 TRANSITIONS BETWEEN POTENTIAL WELLS

In this Section we will examine from a general viewpoint a problem which is frequently met in several applications. The problem is that of the transition between two potential wells.

Let us consider the case described by the equations:

$$\dot{x} = v$$

$$\dot{v} = -\beta v - \frac{dU(x)}{dx} + A(t) \qquad (1.71)$$

where $A(t)$ is, as usual, the white stochastic force. This model describes, for example, Brownian motion in a force field of potential $U(x)$. Let us now suppose that the potential is of the double-well type, shown schematically in Fig. 1.4.

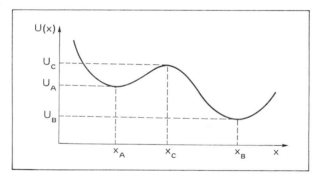

Fig. 1.4 - Double-well potential.

With a potential of this type the above model (Eqs. (1.71)) can also be used to describe chemical reactions, in which case the abscissa represents the reaction coordinate. Such a chemical reaction corresponds to passing from the well of the reactants (A) to that of the products (B).

Problems of this type are also found in other applications, such as flip-flop devices, especially those employing a tunnel diode.

The results obtained will be of general validity and not restricted to one particular physical application, even though we will often refer to Brownian particles jumping from one potential well to another. We will also completely ignore the quantum tunnelling effect.

The Fokker-Planck equation for the transition probability density $p(x, v, t \mid x_0, v_0)$ has the following form:

$$\frac{\partial}{\partial t} p(x, v, t \mid x_0, v_0) + \frac{\partial}{\partial x}(vp) + \frac{\partial}{\partial v}[(-\beta v - (dU/dx))p] = D \frac{\partial^2}{\partial v^2} p \qquad (1.72)$$

It is impossible to analytically solve this equation because of the non-linearity of the potential $U(x)$. We can, however, try to make it more manageable. One initial simplification is possible in the case of high viscosity, where v relaxes much more rapidly than x. By performing a direct adiabatic elimination (i.e. by putting $\dot{v} = 0$ in the second Eq. (1.71)) we obtain:

$$\dot{x} = -\frac{1}{\beta}\frac{dU}{dx} + \frac{1}{\beta}A(t) \tag{1.73}$$

The associated equation for the transition density $p(x, t \mid x_0)$, which in this case is called the Smoluchowski equation, is:

$$\frac{\partial}{\partial t} p(x, t \mid x_0) + \frac{\partial}{\partial x}\left(-\frac{1}{\beta}\frac{dU}{dx} p\right) = \frac{D}{\beta^2}\frac{\partial^2}{\partial x^2} p \tag{1.74}$$

The technique of direct adiabatic elimination is, however, fairly brutal and one can raise serious doubts about the validity of it in the presence of stochastic forces.

Kramers was the first to demonstrate how it is possible to obtain the Smoluchowski equation as a limiting case of the Fokker-Planck equation for the case of high viscosity. Operating directly upon the Fokker-Planck equation, instead of upon the Langevin equation, removes the reservations about the mathematical manipulations.

In what follows we give a brief outline of Kramers' reasoning, which has been taken by other authors who have also calculated successive approximations. Kramers rewrote Eq. (1.72) in a different way, adding and subtracting terms, in such a way as to demonstrate that the quantity $\sigma(x, t \mid x_0)$, defined as

$$\sigma(x, t \mid x_0) = \int_{v + \beta x = \text{const}} p(x, v, t \mid x_0, v_0)\, dv \tag{1.75}$$

exactly satisfies the Smoluchowski equation.

The definition of σ may appear to be arbitrary but, on the contrary, demonstrates the depth of Kramers' physical intuition. In fact, if we had not imposed the condition that

$$v + \beta x = \text{constant} \tag{1.76}$$

then the steps would have been purely algebraic, whereas we know that the reduction to the Smoluchowski equation is only possible in the case of high viscosity. The above condition (Eq. (1.76)) ensures this. In fact, differentiating w.r.t. time, Eq. (1.76) becomes

$$\dot{v} + \beta\dot{x} = 0 \tag{1.77}$$

therefore, since β is high, v has much more rapid dynamics than x.

We have therefore obtained, for the case of high viscosity, the Smoluchowski equation which we can rewrite as

$$\frac{\partial\sigma}{\partial t} + \frac{\partial}{\partial x}\left(-\frac{1}{\beta}\frac{dU}{dx}\sigma\right) = \tilde{D}\frac{\partial^2}{\partial x^2}\sigma \tag{1.78}$$

$$\tilde{D} = D\,\beta^{-2}$$

Now we propose to calculate the transition rate from A to B in the stationary case ($\dot{\sigma} = 0$). The choice of a stationary state implies, in a strict sense, a continuous supply of Brownian particles in A and a continuous extraction of Brownian particles in B. It is possible, however, that we can obtain some indications applicable also to the quasi-stationary state with only slowly varying concentrations. Let us also suppose that the concentration of Brownian particles in A is much greater than that in B.

The stationary state solution $\tilde{\sigma}(x)$ obeys the equation

$$\frac{\partial}{\partial x}\left\{-\frac{1}{\beta}\frac{dU}{dx}\tilde{\sigma}-\tilde{D}\frac{\partial\tilde{\sigma}}{\partial x}\right\}=0$$

so that

$$\tilde{j}=-\frac{1}{\beta}\frac{dU}{dx}\tilde{\sigma}-\tilde{D}\frac{\partial\tilde{\sigma}}{\partial x}=\text{constant} \tag{1.79}$$

Since the Smoluchowski equation has the form of a continuity equation, $\tilde{j}$ can be identified with the probability flow density. Eq. (1.79) can be rewritten in the equivalent form

$$\tilde{j}=-\tilde{D}e^{-U/\beta\tilde{D}}\frac{\partial}{\partial x}\{\tilde{\sigma}\,e^{U/\beta\tilde{D}}\} \tag{1.80}$$

Introducing the fluctuation-dissipation relationship $D=\beta kT$, so that $\tilde{D}=kT/\beta$, this is transformed into:

$$\tilde{j}=-\tilde{D}e^{-U(x)/kT}\frac{\partial}{\partial x}\{\tilde{\sigma}(x)\,e^{U(x)/kT}\} \tag{1.81}$$

Integrating w.r.t. x between two arbitrary points, we have

$$-\tilde{D}\,[\tilde{\sigma}(x)\,e^{U(x)/kT}]_{x=x_\alpha}^{x=x_\beta}=\tilde{j}\int_{x_\alpha}^{x_\beta}e^{U(x)/kT}dx \tag{1.82}$$

In this way it is possible, in principle, to calculate the diffusion current j. However, the integral which appears on the r.h.s. of Eq. (1.82) is not normally calculable analytically.

In the case of the double-well potential, it was Kramers who showed how an approximate calculation could be made.

Supposing that the potential barrier is not too low, the velocity of passing over the barrier will be small with respect to the dynamics of the settling of the particles on the well bottom. Thus Kramers assumed that it was possible to describe the statistical distribution of the particles inside well A as an equilibrium distribution. He also assumed that it was possible to approximate the shape of the potential around the bottom of well A and around the top of the energy barrier C, using quadratic expressions:

$$U_A(x)=U_A+\frac{1}{2}(2\pi\omega_A)^2(x-x_A)^2 \qquad \text{around } A \tag{1.83}$$

$$U_C(x)=U_C-\frac{1}{2}(2\pi\omega_C)^2(x-x_C)^2 \qquad \text{around } C \tag{1.84}$$

On the basis of these hypotheses, the number of particles n_A in the vicinity of A is

$$n_A = \int_{-\infty}^{+\infty} \sigma_A \, e^{-U_A(x)/kT} \, dx \tag{1.85}$$

where $\sigma_A \, dx$ is the number of particles between x_A and $x_A + dx_A$, and extending the limits of the integration is justified by the fact that its dominating contribution comes from a small region around A. Thus n_A can be calculated as a Gaussian integral, resulting in:

$$n_A = \frac{\sigma_A}{\omega_A} \sqrt{\frac{kT}{2\pi}} \, \exp[-U_A/kT] \tag{1.86}$$

Using the same hypotheses, from Eq. (1.82) we obtain

$$\tilde{j} = \tilde{D} \, \frac{\sigma_A}{\displaystyle\int_A^B e^{U(x)/kT} \, dx} \tag{1.87}$$

where we have ignored the integrated term in B since we have assumed that

$$\sigma_A \gg \sigma_B$$

The rate constant of exchange ("Kramers rate" r) is equal to the ratio between the number of particles that jump in one second, $\tilde{j}$, and the number of reacting particles, n_A. It is therefore given by:

$$r = \frac{\tilde{j}}{n_A} = \tilde{D} \, \frac{\omega_A \sqrt{2\pi/kT}}{\displaystyle\int_A^B e^{U(x)/kT} \, dx} \, \exp[U_A/kT] \tag{1.88}$$

Given the form of the integrand which appears on the r.h.s., the dominating contribution comes from around C. Therefore, substituting the approximate expression $U_A(x)$ for $U(x)$, and extending the limits of the integration to infinity, it is also possible to evaluate this quantity as a Gaussian integral. We thus obtain:

$$r = \frac{2\pi\tilde{D}}{kT} \, \omega_A \, \omega_C \, \exp\left[-\frac{U_C - U_A}{kT}\right] = 2\pi \, \frac{\omega_A \, \omega_C}{\beta} \, \exp\left[-\frac{U_C - U_A}{kT}\right] \tag{1.89}$$

It is to be noted that r is a rate constant, whereas the total number of particles passing the barrier in one second is equal to $r \cdot n_A$.

What we have considered so far may be directly applied to chemical reactions, by simply substituting the number of particles in A and in B, respectively, with reactant and product concentrations. In this context, Eq. (1.89) for the rate constant is the well-known Arrhenius equation.

All of this, however, refers to the case of high viscosity. It is also possible to treat in a similar fashion the limiting case of low viscosity, even though in this case the rate is no longer a fast-changing variable with respect to the position.

We will limit ourselves here to an outline of the method of calculation, considering the case of potentials which give rise to an oscillatory motion. One changes the variables in the Fokker-Planck equation, introducing an energy term. In the low viscosity case the energy varies very little over one period, and so it can take the role of a slowly changing variable. One thus obtains a Smoluchowski equation for the energy which, when treated in the same manner as the preceding case, gives the following result:

$$r = \beta \ \frac{U_C - U_A}{kT} \ \exp\left[-\ \frac{U_C - U_A}{kT}\right] \tag{1.90}$$

One can see that here r is proportional to β, whereas in the high viscosity case r is proportional to $1/\beta$. In intermediate cases more refined analytical or numerical techniques are required. The behaviour of r as a function of the reciprocal of the friction coefficient is represented qualitatively in Fig. 1.5.

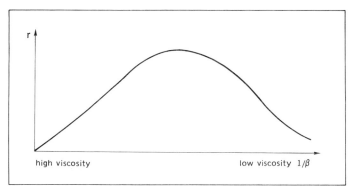

Fig. 1.5 - **Qualitative behaviour of the rate constant for the barrier crossing process plotted against the reciprocal of the friction coefficient.**

1.7 THE EHRENFEST URN

The "Langevin" approach, which we have so far described, can be defined, in a certain sense, as an individual approach: we have, in fact, considered the force acting upon a single Brownian particle, for which we have given a stochastic description. The Fokker-Planck equation, which describes the temporal evolution of the probability density, that is the modifications in the distribution of a particle population, was introduced by us as a means of resolving the problem formulated in terms of the Langevin equation.

This method of approach is particularly useful and convenient when there is a mesoscopic "indicator", such as the Brownian particle, so that it is possible to formulate reasonable hypotheses on the laws of motion.

It can also be seen that it is not even necessary to operate with a force in the strict sense of the word. For example, as we shall see later, when dealing with kinetic theory and the termodynamics of irreversible processes we can write equations of the following kind:

$$\dot{z}(t) = f(z(t))$$

where z indicates an appropriate set of macroscopic variables and the $f(z)$ are generalised forces. These equations are susceptible to a stochastic extension so that they are considered as the deterministic limit of Langevin-type equations:

$$\dot{z}(t) = f(z(t)) + A(t)$$

In other cases a collective approach is preferable, in which one begins directly from a time-evolution equation for the probability distribution of a population of objects (the master equation).

Before going on to discuss this important subject, it is advisable to consider a simple model which will help to make the following discussion more concrete and intuitive. We will consider the model first proposed by Ehrenfest for discussing the problem of approaching equilibrium.

Let us take a system composed of two identical boxes and of $2N$ numbered balls, each of which is to be found in one or the other of the two boxes. The operation consists in finding a random integer between 1 and $2N$, and in moving the ball carrying that number from its original box to the other one. The next operation takes place in the same manner and it is even possible that the same ball will again be extracted and moved.

Let us now suppose that in the initial condition all the balls are in box 1. The next operation will certainly be the moving of a ball to box 2. The following operation could be the return of that ball to box 1, but it is obviously much more probable that another ball will be moved to box 2. Continuing in this way a transfer from the fuller box to the other will always be statistically preferred, and the system will tend towards a state of equilibrium in which both boxes contain the same number of balls. Obviously, fluctuations are possible both during the approach towards equilibrium and around the equilibrium point. The change in the value of N_1 (the number of balls to be found in box 1) with time is shown qualitatively in Fig. 1.6, in the continuous time approximation.

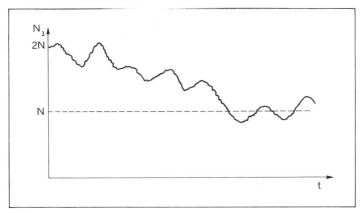

Fig. 1.6 -Time evolution of the variable N_1 in the Ehrenfest model.

We can give a microscopic description of our system, at every step, by specifying in which box every one of the $2N$ balls can be found. Otherwise, if we ignore the difference between one ball and another, we can give a macroscopic description, specifying only the numbers of balls N_1 and N_2 to be found in each of the boxes. Besides, as $N_1 + N_2 = 2N$, only one macroscopic variable is necessary for a description of our system.

Let us now consider the stochastic process constituted by the family of temporal histories $N_1(k \Delta t)$, where Δt is the interval, taken as constant, between one operation and the next. This is a Markovian process, since the state of the system at the k−th instant, $N_1(k \Delta t)$, and the $(k + 1)$−th operation (that is independent of previous history) are sufficient to determine the successive state $N_1((k + 1) \Delta t)$.

We will now write the time-evolution equation for the transition density $p(N_1, s \Delta t \mid N_{01})$, which is the probability of finding N_1 balls in box 1 at time $s \Delta t$, given that there were N_{01} at $t = 0$. For the sake of simplicity we will put $\Delta t = 1$ and consider N_{01} fixed; we can then write $p(N_1, s)$ in place of $p(N_1, s \Delta t \mid N_{01})$.

The probability of finding N_1 balls at time $s + 1$ comes from the sum of two terms:

- the contribution deriving from a transfer from box 2 to box 1, given the value $N_1 - 1$ at time s;
- the contribution deriving from a transfer from box 1 to box 2, given the value $N_1 + 1$ at time s.

The former contribution results from the product of the probability of having $N_1 - 1$ balls in box 1 at time s with the probability that a transfer from box 2 to box 1 will occur. This is equal to the fraction of the balls in box 2 at time s, $[2N - (N_1 - 1)]/2N$. The latter contribution is given by the product of the probability of finding $N_1 + 1$ balls in box 1 at time s by the probability that we have a transfer from box 1 to box 2, which amounts to $(N_1 + 1)/2N$. Thus:

$$p(N_1, s+1) = p(N_1 - 1, s)[1 - (N_1 - 1)/2N] + p(N_1 + 1, s)(N_1 + 1)/2N \qquad (1.91)$$

It is convenient to rewrite this equation using a new macroscopic variable n (order parameter), equal to one half of the difference between N_1 and N_2:

$$n = (N_1 - N_2)/2 \qquad (1.92)$$

The time evolution equation (Eq. (1.91)) then becomes:

$$p(n, s+1) = \frac{N - (n-1)}{2N} p(n-1, s) + \frac{N + (n+1)}{2N} p(n+1, s) \qquad (1.93)$$

This is our basic equation. It is worth noting that, without using anything like the Langevin equation, we have calculated directly the transition velocities between macroscopic states characterised by different values of the order parameter (collective approach).

Supposing that Δt is small with respect to the time scale in which we are interested, we can put (recalling that $\Delta t = 1$):

$$p(n, s+1) - p(n, s) \simeq \frac{\partial}{\partial s} p(n, s) \qquad (1.94)$$

Subtracting $p(n, s)$ from both sides of Eq. (1.93), we have:

$$\frac{\partial}{\partial s} p(n, s) = \frac{N - (n-1)}{2N} p(n-1, s) + \frac{N + (n+1)}{2N} p(n+1, s) - p(n, s) \qquad (1.95)$$

This is the first example we have met here of a master equation: it has the form of a balance equation, and renders explicit the fact that the net rate of change in the probability density results from the competition between "birth" terms (transitions such as $(n-1) \to n$ and $(n+1) \to n$) and "death" terms (transitions such as $n \to (n+1)$ and $n \to (n-1)$).

Let us now perform the continuous limit also for the variable n, assuming that the total number of balls, $2N$, is sufficiently large to allow this approximation. Rewriting the r.h.s. of the master equation (Eq. (1.95)) in the form:

$$\frac{N - (n-1)}{2N} [p(n-1, s) - p(n, s)] +$$

$$+ \frac{N + (n+1)}{2N} [p(n+1, s) - p(n, s)] + \frac{1}{N} p(n, s) =$$

$$= \frac{1}{2} [p(n-1, s) + p(n+1, s)] + \frac{n}{2N} [p(n+1, s) -$$

$$-p(n, s) + p(n, s) - p(n - 1, s)] + \frac{1}{2N} [p(n - 1, s) -$$

$$-p(n, s) + p(n + 1, s) - p(n, s) + 2p(n, s)]$$

and noting that in the continuous limit:

$$p(n + 1, s) - 2p(n, s) + p(n - 1, s) \rightarrow \frac{\partial^2}{\partial n^2} p(n, s)$$

$$p(n + 1, s) - p(n, s) \rightarrow \frac{\partial}{\partial n} p(n, s)$$

(1.96)

we obtain:

$$\frac{\partial}{\partial s} p(n, s) = \frac{1}{N} \frac{\partial}{\partial n} [n \, p(n, s)] + \frac{1}{2} \frac{\partial^2}{\partial n^2} p(n, s)$$

(1.97)

which is clearly an equation of Fokker-Planck type. If it is desired to introduce time in place of the dimensionless variable s, it is sufficient to remember that if

$$t = s\Delta t$$

then

$$\frac{\partial}{\partial s} = \Delta t \frac{\partial}{\partial t}$$

so that Eq. (1.97) becomes

$$\frac{\partial p}{\partial t} = \frac{\alpha}{2} \frac{\partial^2 p}{\partial n^2} + \frac{\alpha}{N} \frac{\partial}{\partial n} (np)$$

where $\alpha = 1/\Delta t$.

We have thus derived an important result: the Fokker-Planck equation is a limiting case of the master equation. We will later see that this result is not limited only to this example.

The results obtained so far have been obtained within the framework that we have called the collective approach: we have studied directly the time evolution of the probability distribution of the population of $2N$ balls, and we have seen that, as a function of the different approximations we have made, it can be described by three different equations:

—   the discrete equation (Eq.(1.93)). which in this case is exact;
—   the master equation (Eq. (1.95)), which is obtained from the former in the continuous limit of the temporal variable, and that has a discrete state space;
—   the Fokker-Planck equation (Eq.(1.97)), which is obtained from Eq. (1.95) in the limit at which n can be considered as continuous, and is therefore a master equation with continuous state space. The appearance of the Fokker-Planck equation naturally raises the question as to whether or not a Langevin-type approach to the problem of Ehrenfest is possible. In this case there is no sense in raising the problem of the force acting upon a single particle. However, the answer to the question is positive. To understand why this is possible, we first note that the deterministic part of the Langevin equation coincides with the time-evolution equation for average values. In fact, averaging the equation

$$\dot{v} = -\beta v + A(t)$$

and observing that the ensemble averaging operation commutes with the derivative w.r.t. time, we obtain the time-evolution law for $\langle v(t) \rangle$:

$$\frac{\partial}{\partial t} \langle v \rangle = -\beta \langle v \rangle$$

As we have seen, the basic reasoning behind the Langevin equation consists in adding a fluctuating force to a deterministic equation valid for average values. We can now think about applying this line of reasoning to our case. First of all, we must find the deterministic time-evolution law for the average value of the order parameter n or, equivalently, of N_1. We see that

$$\frac{d \langle N_1 \rangle}{dt} = \left\{ \begin{array}{c} \text{average velocity} \\ \text{of the transition} \\ 2 \to 1 \end{array} \right\} - \left\{ \begin{array}{c} \text{average velocity} \\ \text{of the transition} \\ 1 \to 2 \end{array} \right\}$$

and therefore

$$\frac{d \langle N_1 \rangle}{dt} = k \left[\frac{\langle N_2 \rangle}{2N} - \frac{\langle N_1 \rangle}{2N} \right] = k \left[1 - \frac{\langle N_1 \rangle}{N} \right]$$

With a simple change of variables we get:

$$\frac{d \langle n \rangle}{dt} = -k \langle n \rangle / N \tag{1.98}$$

which has the solution (corresponding to the initial condition $n(0) = N$):

$$\langle n(t) \rangle = N \exp(-kt/N)$$

Eq. (1.98) can be transformed into a Langevin equation by adding a stochastic white force:

$$dn(t) = -(k/N) n\, dt + dw(t) \tag{1.99}$$

which is associated with the Fokker-Planck equation

$$\frac{\partial}{\partial t} p(n, t \mid n_0) = (k/N) \frac{\partial}{\partial n}(np) + (\sigma^2/2) \frac{\partial^2}{\partial n^2} p$$

which coincides with the equation obtained as the continuous limit of the master equation, given that $k = \alpha$, $\sigma^2 = \alpha$. This result justifies *a posteriori* the addition of the stochastic white force to the time-evolution equation for average values. Besides, as the Langevin equation is equivalent to the Fokker-Planck equation, this result also shows the limits of validity for such a procedure.

1.8 THE MASTER EQUATION

We will now go into the important question of the master equation, of which we have seen an example in the preceding paragraph. For its general derivation, we can consider a homogeneous Markovian process, for which the Chapman-Kolmogorov equation (1.37) can be written, save for minor modifications of the symbols, as

$$p(y, t + s \mid z) = \int p(y, s \mid x) p(x, t \mid z)\, dx \tag{1.100}$$

When s is very small, it is reasonable to suppose that the transition density differs from a Dirac delta for terms that are at most of order s. Thus:

$$p(y, s \mid x) = [1 - a(x)s] \delta(y - x) + W(y \mid x)s + o(s) \tag{1.101}$$

where

$$\lim_{s \to 0} o(s)/s = 0$$

In this equation $W(y \mid x)$ is the transition density per unit time, whereas $a(x)$ is the probability of a transition occurring from x in the interval s. Therefore:

$$a(x) = \int W(y \mid x) \, dy \tag{1.102}$$

Substituting Eq. (1.101) in the Chapman-Kolmogorov equation, we obtain:

$$p(y, t + s \mid z) = \int [(1 - a(x)s) \, \delta(y - x) + W(y \mid x) s] \, p(x, t \mid z) \, dx =$$

$$= [1 - a(y)s] \, p(y, t \mid z) + s \int W(y \mid x) p(x, t \mid z) \, dx$$

Going to the limit where s tends to zero, this formula becomes:

$$\frac{\partial}{\partial t} p(y, t \mid z) = -a(y)p(y, t \mid z) + \int W(y \mid x) p(x, t \mid z) \, dx$$

Incorporating Eq. (1.102) gives us the master equation:

$$\frac{\partial}{\partial t} p(y, t \mid z) = \int [-W(x \mid y)p(y, t \mid z) + W(y \mid x)p(x, t \mid z)] \, dx \tag{1.103}$$

This clearly has the form of a balance equation: the rate of variation of $p(y, t \mid z)$ is determined by the competition between a positive term (birth term), due to the transitions arriving in y, and a negative term (death term), due to the transitions leaving y.

The master equation is an extremely important conceptual instrument in the theory of stochastic processes. It becomes particularly useful where it is possible to formulate hypotheses on the rate of transition. In these cases the master equation is the point of departure for the development of the mathematical treatment without having to go through the "individual" approach à la Langevin. In the case of the Ehrenfest urn we have proceeded precisely in this manner.

As suggested by our example, we can consider the Fokker-Planck equation as a particular case of the master equation characterised by a certain structure on the r.h.s. of the equation. It is possible to carry out a series expansion, in a systematic way, of the master equation (the so-called van Kampen Ω- expansion), and to demonstrate under which conditions one obtains the Fokker-Planck equation from the first terms of the expansion. We will not go into the argument here, it being sufficient to have shown the plausibility of this approach for the case of the Ehrenfest model. The reader interested in this topic is referred to van Kampen's book.

In conclusion, we will write the form of the master equation for the case where the process is one of discrete states, or rather, where the variable of interest can only take discrete values and does not have a continuous range. In perfect analogy with the continuous form of the master equation (Eq. (1.103)), we obtain:

$$\frac{d}{dt} p_n(t) = \sum_k [W_{kn} p_k(t) - W_{nk} p_n(t)]$$

(1.104)

2

Applications of the Theory of Markovian Processes

2.1 JOHNSON NOISE AND THE NYQUIST THEOREM

The microscopic physical basis of Ohm's law lies mainly in the fact that the innumerable interactions between the conduction electrons and lattice impurities within a conducting material can be schematized, at a macroscopic level, as a frictional force. This force causes the current, which plays the role of order parameter (proportional to the average velocity of the charge carriers in the direction of the electric field), to be directly proportional to the applied voltage.

It is to be stressed, however, that in this case the possibility of describing the dynamic behaviour of the electrons as a "Brownian motion" does not derive so much from their smaller mass with respect to the particles composing the "viscous medium" (the crystal lattice and its impurities) as from the fact that the average energy transferred from the lattice to the charge carriers during every interaction is much less than their average kinetic energy.

In other words, we can say that electronic circuit equations only demonstrate the "dissipating" effect of the innumerable interactions taking place. Anyway, as we have seen above, every dissipative effect must be linked to a "fluctuation" effect. It is therefore logical to suppose that such a stochastic effect can also take place within electronic devices.

In fact, the effects of thermal noise were first observed by Johnson in 1927. The explanation of his observations is given by the Nyquist theorem which asserts that, over a frequency interval Δf, the mean square potential $\overline{V^2}$ across a circuit of internal resistance R at a temperature T is given by:

$$\overline{V^2} = 4R k T \Delta f \tag{2.1}$$

Assuming $C(\tau) = \langle V(t) V(t + \tau) \rangle$, the function

$$G(f) = 4 \int_0^\infty C(\tau) \cos(2 \pi f \tau) d\tau \tag{2.2}$$

introduced in Chapter 1 is also referred to as the spectral density since it represents the ensemble average of the power dissipated by an electric circuit of unit resistance subject to thermal fluctuations of the voltage whose Fourier components have frequencies between f and $f + df$.

Thus, as the power dissipated by a circuit of resistance R under a potential difference V is given by V^2/R, the definition (2.2) allows us to rewrite the Nyquist theorem in the following form which is often used in the literature:

$$G(f) = 4kT \qquad (2.3)$$

We will now attempt a microscopic derivation of the Nyquist theorem. We will consider a simple electric circuit consisting of one resistance in thermal equilibrium at a temperature T. Let us also suppose that, as mentioned above, the thermal "motion" of the electrons within the circuit can be described, in tems of classical mechanics, by the following Langevin equation:

$$m \, du/dt = -(m/t_c)\, u + \tilde{F}(t) \qquad (2.4)$$

where u is the electron velocity, $\tilde{F}(t)$ is a white noise term and t_c represents the average period during which the interactions between the lattice impurities and the charge carriers destroy the memory of their preceding dynamic state.

From Eq. (2.4) we can therefore obtain the correlation function for the charge carrier velocity:

$$C(\tau) = \langle u(t)\, u(t+\tau) \rangle \qquad (2.5)$$

On substituting, in this equation, the formal solution of the Langevin equation

$$u(t) = \int_0^t \exp[-\beta(t-s)] \tilde{F}'(s)\, ds + u_0 \exp(-\beta t) \qquad (2.6)$$

where $\beta = 1/t_c$ and $\tilde{F}'(s) = \tilde{F}(s)/m$, we obtain:

$$C(\tau) = \int_0^t \int_0^{t+\tau} \exp[-\beta(t-s)] \exp[-\beta(t+\tau-s')] \cdot$$

$$\cdot \langle \tilde{F}(s)\, \tilde{F}(s') \rangle \, ds\, ds' + u_0^2 \exp[-\beta(2t+\tau)] \qquad (2.7)$$

Now, taking account of the statistical properties of the fluctuating force, we obtain:

$$C(\tau) = \sigma^2 \int_0^t \int_0^{t+\tau} \exp[-\beta(t-s) - \beta(t+\tau-s')] \cdot$$

$$\cdot \delta(s-s')\, ds\, ds' + u_0^2 \exp[-\beta(2t+\tau)]$$

which, recalling the properties of the Dirac function, simplifies to

$$C(\tau) = (u_0^2 - \sigma^2/2\beta) \exp[-\beta(2t+\tau)] + (\sigma^2/2\beta) \exp(-\beta\tau) \qquad (2.8)$$

For a metal at room temperature, t_c is of the order of magnitude of 10^{-13} seconds, and we can thus ignore the first term on the r.h.s. of the above equation since it depends on the macroscopic time t, obtaining:

$$C(\tau) = \bar{u}^2 \exp(-\beta\tau) \qquad (2.9)$$

where $\bar{u}^2 = \sigma^2/2\beta = \sigma^2 t_c/2$.

This equation allows us to obtain the correlation function of the voltage across the resistance. Let us rewrite the potential difference V, measured across the resistance (which we assume to have unit length and unit cross section), as:

$$V = RI = R\, e \sum_i u_i \qquad (2.10)$$

where I is the current flowing through the circuit, e is the electron charge and u_i is the average velocity acquired by the i-th electron due to the applied potential.

On taking into account the equation previously obtained for the correlation function of the electronic velocity, we therefore have:

$$C(\tau) = \langle V(t)\, V(t + \tau) \rangle = N(e\,R)^2 \bar{u}^2 \exp\left(-\frac{\tau}{t_c}\right) \qquad (2.11)$$

where N is the number of electrons per unit volume.

Consequently, Eq. (2.2) yields

$$G(f) = 4N \int_0^\infty (e\,R)^2\, \bar{u}^2\, \exp(-\tau/t_c) \cos(2\pi f \tau)\, d\tau =$$
$$= 4N\,[(e\,R)^2\, \bar{u}^2\,]\, t_c/[1 + (2\pi f t_c)^2\,] \qquad (2.12)$$

As stated previously, for a metal at room temperature, t_c is of the order of magnitude of 10^{-13} seconds and if we do not take into consideration signals at extremely high frequencies, $2\pi f t_c \ll 1$.

Eq. (2.12) can therefore be written approximately as:

$$G(f) \cong 4N(eR\bar{u})^2\, t_c \qquad (2.13)$$

Now, let us make the hypothesis that the statistics of the electrons at thermal equilibrium can be described by a Maxwellian distribution function. Later it will be shown that the voltage fluctuations deriving from thermal noise turn out to be independent of the specific details of the statistics obeyed by the charge carriers. The choice of the Maxwell distribution function, therefore, does not detract from the general nature of what has been demonstrated here. Under this assumption we have, from the theorem of equipartition of energy:

$$m\,\bar{u}^2\,/2 = kT/2 \qquad (2.14)$$

and therefore:

$$\bar{u}^2 = kT/m \qquad (2.15)$$

Consequently, we have:

$$\bar{V}^2 = G(f)\,df = 4N\,(e\,R)^2\ (kT/m)\ t_c\,df \qquad (2.16)$$

Let us remark that, having assumed unitary dimensions, the resistance R equals $1/(N e^2\, t_c\,/m)$. Therefore the preceding equation can be rewritten:

$$\bar{V}^2 = 4kTR\,df \qquad (2.17)$$

which is exactly the Nyquist relationship that we wanted to derive. In the reasoning above we have assumed that the distribution function of the electrons in the circuit was Maxwellian. This has allowed us to use the theorem of equipartition of energy in order to obtain the dependence of u on the temperature T.

We will now show that having assumed a Maxwellian distribution for the electron velocities does not reduce the validity of our final expression for the power spectrum. To this end, let us consider a circuit consisting of two resistances, R_A and R_B, wired in series under thermal equilibrium at a temperature T.

Let us then imagine that within resistance R_A the electrons follow Fermi-Dirac statistics, whereas within resistance R_B we have "Maxwellian" charge carriers. The latter resistor thus generates a thermal noise $\bar{V}_B^2 = 4R_B kT df$ which is associated with a current noise equal to:

$$\bar{I}_B^2 = \bar{V}_B^2 / (R_A + R_B)^2 \tag{2.18}$$

Therefore, the power transferred from R_B to R_A is given by:

$$\bar{V}_B^2 \, R_A / (R_A + R_B) \tag{2.19}$$

An analogous reasoning tells us that the power transferred from resistor R_A to R_B is equal to

$$\bar{V}_A^2 R_B / (R_A + R_B) \tag{2.20}$$

where $\bar{V}_A^2$ is the thermal noise generated by "quantum" electrons.

As we have assumed that both of the resistances are at the same temperature, from the second law of thermodynamics heat can not flow from one resistance to the other. It follows that:

$$\bar{V}_A^2 / R_A = \bar{V}_B^2 / R_B = 4 \, k \, T \, df \tag{2.21}$$

Thus, we have demonstrated that the Nyquist theorem, as given by Eq. (2.17), is independent of the details of the statistics to which the charge carriers are subject.

2.2 BROADENING OF SPECTRAL LINES

One of the earliest and most powerful methods of investigating the internal structures of atoms and molecules was the study of their absorption and emission spectra.

Quantum theory then allows us to relate the spectral distribution of the radiation intensity, $I(f)$ (also known as the shape function), to the number of transitions that take place, in unit time, between the allowed quantum levels. As these transitions can only take place at well-defined frequencies ω_1, ω_2, ω_3 ... the emitted energy for every other frequency is zero. We can therefore assume that, in general, the shape function $I(\omega)$, which is characteristic of each particular substance, takes the form of a collection of Dirac deltas.

The first spectral observations, in fact, demonstrated the excellent agreement between the experimental distributions and the predictions of the quantum theory.

However, by increasing the resolution of the instrument, it can be seen that the various spectral lines are not extremely sharp, and that, in reality, they consist of a more complicated structure about the central frequency predicted by elementary quantum theory applied to the case of single atoms or molecules.

Various mechanisms have been proposed, over the years, for explaining this broadening phenomenon. Let us now examine, for the case where the sample is in the gas phase, an effect which is more strictly related to the statistical mechanics of gases; i.e. collision broadening.

When, in fact, two gas molecules find themselves very close to each other, the interaction between their respective electronic clouds causes changes in their energy levels so that the transition, corresponding to the frequency ω_0 in the case of the isolated molecule, now occurs at a new frequency $\omega_0 + d\omega_0$.

Here, and in what follows, we shall assume that the molecular motion is quasi-classical, so that we can use the concept of particle trajectory. We also assume that the interaction times are extremely small (classical collision) and that the interaction between the electronic clouds does not change the pre-collision quantum state.

Let us now try to formulate, in a more quantitative manner, that which has been briefly outlined above.

Within classical theory, the emission of radiation of a frequency ω_0 by an excited atom or molecule can be interpreted as a process of irradiation due to the oscillatory motion of an electric charge. Therefore let

$$x(t) = \exp(i\,\omega_0 t) \tag{2.22}$$

be the function that describes the temporal behaviour of the atomic oscillator. In this case, ω_0 represents the frequency in the complete absence of perturbation.

The effect of the electrostatic interaction due to collisions with other molecules can, therefore, be summarised with the addition of a perturbative term $\eta(t)$ in the phase of the oscillator. In this way, the solution of the equation of our harmonic oscillator becomes:

$$x(t) = \exp[i\omega_0 t + i\,\eta(t)] \tag{2.23}$$

Within the model reported here, the broadening of a spectral line due to collisions between gas molecules is related to the perturbation of the perfectly periodic motion of the isolated atomic oscillator.

The shape function for the spectral lines $I(\omega)$ is defined as the squared modulus of the Fourier transform of $x(t)$:

$$I(\omega) = \lim_{T \to \infty} \left| \frac{1}{2\pi T} \int_{-T/2}^{T/2} x(t)\,\exp(-i\omega t)\,dt \right|^2 = \tag{2.24}$$

$$= \lim_{T \to \infty} (1/2\pi T) \left| \int_{-T/2}^{T/2} \exp[-i(\omega - \omega_0)t + i\,\eta(t)]\,dt \right|^2 \tag{2.25}$$

In what follows, in order not to complicate the writing of the equations, we will assume that the frequencies are measured starting from the unperturbed frequency, ω_0. Then the previous equation becomes:

$$I(\omega) = \lim_{T \to \infty} (1/2\pi T) \left| \int_{-T/2}^{T/2} \exp[-i\omega t + i\,\eta(t)]\,dt \right|^2 \tag{2.26}$$

Collision between molecules is a random phenomenon and, consequently, $y(t) = \exp[i\,\eta(t)]$ is also a stochastic process. Generally, however, in the broadening of spectral lines, conditions such as temperature, pressure, etc. are considered to be stationary, so that the statistical distribution of the collisions with respect to time is constant. It follows, therefore, that $x(t)$ is also a stationary stochastic process and, on changing the limits of integration, Eq. (2.26) can be rewritten as:

$$I(\omega) = (1/\pi)\,Re\left[\int_0^\infty \phi(s)\,\exp(-i\omega s)\,ds \right] \tag{2.27}$$

where

$$\phi(t) = \lim_{T \to \infty} (1/T) \int_{-T/2}^{T/2} x^*(t)\,x(t+s)\,ds = \overline{x^*(t)\,x(t+s)} \tag{2.28}$$

is the correlation function of the stochastic oscillations. Given the stationary nature of the process, we can substitute the time average which appears in the preceding equation with the ensemble average:

$$\phi(t) = \langle x^*(0)\,x(t) \rangle = \langle \exp[i\,\eta\,(t)] \rangle \tag{2.29}$$

Let us now try to obtain an explicit equation for $\phi(t)$. Thus, let

$$\Delta\phi = \phi(t + \Delta t) - \phi(t) \tag{2.30}$$

be the change in the correlation function $\phi(t)$ over the infinitesimal period of time Δt. Recalling Eq. (2.29) we have:

$$\Delta\phi = \langle \exp[i\,\eta(t + \Delta t)]\rangle - \langle \exp[i\,\eta(t)]\rangle =$$
$$= \langle \exp[i\,\eta(t)]\exp(i\,\Delta\eta)\rangle - \langle \exp[i\,\eta(t)]\rangle \tag{2.31}$$

where $\Delta\eta$ is the change of phase of the electronic oscillations due to the collisions which took place during the interval Δt. Then, since we assume an extremely short period of interaction between the molecules, the phase shift, $\Delta\eta$, becomes independent of $\eta(t)$ and consequently we can rewrite the previous equation as:

$$\Delta\phi = \langle \exp[i\,\eta(t)]\rangle\,[\langle \exp(i\,\Delta\eta)\rangle - 1] = -\phi(t)\langle 1 - \exp(i\,\Delta\eta)\rangle \tag{2.32}$$

We will now use $P(b, v)\,db\,dv$ to represent the number of collisions per second between molecules having a relative velocity v and collision parameter b. It must be recalled that the collision parameter is defined as the distance between the lines on which the initial velocity vectors of the two incident particles are to be found. Consequently, we can express the term $\langle 1 - \exp(i\,\Delta\eta)\rangle$ as:

$$\langle 1 - \exp(i\,\Delta\eta)\rangle = \Delta t \iint \{1 - \exp[i\,\eta'(b, v)]\}\,P(b, v)\,db\,dv = \Theta\,\Delta t \tag{2.33}$$

where $\eta'(b, v)$ indicates the phase shift produced by a collision having the parameters b and v Eq. (2.32) thus becomes:

$$\Delta\phi = -\Theta\,\phi\,\Delta t$$

which, in the limit $\Delta t \to 0$, becomes:

$$d\phi/dt = -\Theta\,\phi \tag{2.34}$$

having as its solution:

$$\phi(t) = \exp(-\Theta t) \tag{2.35}$$

On substituting this result in Eq. (2.27), we obtain the line shape function $I(\omega)$:

$$I(\omega) = (\gamma/2\pi)\,\{1/[(\omega - \Delta)^2 + (\gamma/2)^2]\} \tag{2.36}$$

where γ and Δ are parameters which depend upon Θ.

The curve which represents Eq. (2.36) is known as a Lorentzian curve and it describes, in a sufficiently accurate manner, the experimentally observed broadening of spectral lines. Its shape, as can be seen from Fig. 2.1, is similar to that of a Gaussian curve although, for values of $\omega \gg \Delta$, it falls to zero in a more gradual manner. The full-width half-maximum γ of the line can be fairly easily related to physically significant macroscopic quantities. In fact, developing Eq. (2.33) in a more explicit manner we obtain:

$$\gamma = 2N\langle v\,2\pi \int_0^\infty \{1 - \cos[\eta(b,v)]\}b\,db\,\rangle \tag{2.37}$$

where N is the molecular density of the gas. Assuming that the phase shift induced by the collisions is statistically independent of v, we can rewrite the previous equation as:

$$\gamma = 2N \langle v \rangle \langle 2\pi \int_0^\infty [1 - \cos(\eta)] \, b \, db \rangle .$$ (2.38)

Now, from the kinetic theory of gases,

$$N = P/kT , \qquad \langle v \rangle = \sqrt{8\,kT/\pi\,m}$$ (2.39)

where $\langle v \rangle$ is the mean relative velocity, and consequently

$$\gamma = CP/\sqrt{T}$$ (2.40)

where C is a suitable constant. Experimental data do, in fact, confirm this dependence of the full-width half-maximum of spectral lines upon the pressure and temperature as shown by Eq. (2.40).

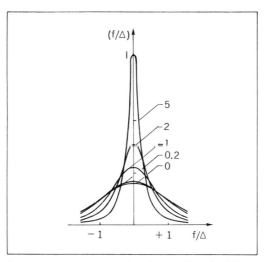

Fig. 2.1 - Lorentzian curves plotted against $f - \Delta$ for different values of the parameter γ/Δ.

In a certain sense we can consider the mechanism studied for the broadening of spectral lines as being closely related to Brownian motion, because also in this case we are concerned with the effects of intermolecular collisions. In any case, it is worth stressing several new and peculiar aspects which the example given here presents.

For Brownian motion, noise has a direct effect upon the classical equation of motion of the particles which manifests itself as an additive stochastic term in the Newton equation. In other words, for Brownian motion, the intermolecular collisions affect the external degrees of freedom of the particle.

In any case, the interaction between the electronic clouds also affects the internal degrees of freedom of the molecules which, as we have seen above, also produces a broadening of the spectral lines. In this case, however, an even more important fact arises, i.e. the noise is no longer additive but multiplicative. In fact, from our reasoning so far, if the unperturbed harmonic oscillator equation is

$$\ddot{x} = -\omega_0^2 x$$ (2.41)

the effect of the intermolecular collisions has been schematized here by introducing a stochastic term in the frequency of the oscillations:

$$\ddot{x} = -(\omega_0 + \eta)^2 x \qquad (2.42)$$

The noise term, therefore, multiplies the variable of interest. Multiplicative noise plays a determining role in many other situations of physical interest.

2.3 THE TUNNEL DIODE

The atoms of a crystalline material occupy clearly defined positions in an ordered structure. Theoretical calculations show that, due to the overlapping of the various electronic wavefunctions, the original atomic energy levels are concentrated into bands.

From the Pauli exclusion principle each band can contain a well defined number of electrons. The importance of all this resides in the fact that the band structure of a solid determines many of its electrical properties. For example, no electrical conduction can occur with only a completely empty band and neither with only a completely filled band. In the latter case, in fact, the motion of the electrons in one direction is exactly compensated by that in the opposite direction and the total charge population in the band does not present any net motion.

Thus it is evident that a metal is a material with a partially filled energy band; an insulating material, however, has only completely filled or completely empty bands separated by broad prohibited energy gaps; finally, a semiconductor has a band structure very similar to that of an insulator, except for the fact that within one of the bands, not completely filled, mobile charge carriers are present. These charge carriers can be introduced by impurities due to atoms having a valence which is different from that of the host species (and occupying equivalent lattice positions), by crystal defects, or even by thermal excitation of electrons from the highest completely-filled band (valence band) to the lowest unoccupied band (conduction band). In fact, in a semiconductor, the energy gap between the bands is less than that between the corresponding bands in an insulator

In order to describe in a correct manner the complete behaviour of charge carriers within a material it is necessary to take into account the Pauli exclusion principle and therefore to use the Fermi-Dirac statistics

$$f(E) = 1/\{\exp[(E - E_F)/k\,T] + 1\} \qquad (2.43)$$

where E_F is called the Fermi energy or Fermi level.

It can also be shown that the Fermi level coincides with the chemical potential of the material in equilibrium. This fact plays an important role in the study, for example, of the physics of solid state devices. In fact, the majority of these depend upon the junctions between different materials and the fact that the Fermi level, at equilibrium, is constant throughout the device constitutes, in many cases, a useful reference and point of departure for the development of a physical theory of the operation of such electronic components.

We will now examine one of these devices, the tunnel diode, in a little more detail.

The tunnel diode is a device consisting of a junction between two highly-doped materials (impurity concentration of the order of 10^{18} cm^{-3}), one with electron acceptor impurities (p material) and the other with electron donor impurities (n material). Because of the high percentage of impurities, the Fermi level, the position of which within the material is related to the number of fixed and mobile charges present, is to be found, in the p region, within the valence band, and in the n region, within the conduction band. In this case both regions are termed degenerate regions.

Let us now try to give a qualitative description of the behaviour of the tunnel diode. First, let us consider the case where there is an external applied voltage. As stated above, because the

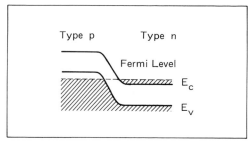

Fig. 2.2 - **Band structure of a tunnel diode in the absence of an external applied voltage. E_v and E_c designate respectively the upper limit of the valence band and the lower limit of the conduction band.**

regions constituting the *p-n* junction are degenerate there is a partial energy overlap of the valence band of the *p* region with the conduction band of the *n* region. Also, it can be shown that because of the high dopant level the potential barrier, to which the charge carriers are subject in the junction region, is quite thin, so that there is a high tunnelling probability. The net flow of electrons in one direction or the other depends not only upon the probability of the transition but also upon the number of charge carriers on each side of the junction and the availability of states having the same energy on the other side of the junction.

We can take it that the "tunnel" current across the junction consists of two components of opposite sign: the Zener current, due to the electrons passing from the valence band of the *p* region to the conduction band of the *n* region, and the Esaki current, due to the flow in the opposite direction. It is clear that, in the absence of an external applied voltage, these currents must compensate one another. Let us suppose that a low direct biasing voltage (i.e. when the positive terminal is attached to the *p* region) is applied across the diode. From the point of view of the structure of the energy bands within the material, the application of this external voltage has the effect of reducing the difference between the upper limit of the valence band of the *p* material and the lower limit of the conduction band of the *n* material. Consequently, there is an increase in the number of available energy states in the *p* material having energies equal to those of the electrons in the conduction band of the *n* material. There is thus an increase in the flow of charge carriers from the *n* region to the *p* region (the Esaki current), while the flow of electrons in the opposite direction, from the *p* region to the *n* region, decreases. In this way, the flow of charge carriers is no longer balanced and there is a consequent increase in the current flowing from the *p* region to the *n* region (direct current). On increasing the direct biasing voltage, however, the situation changes, as there is a decrease in the number of available states in the *p* material with the same energy level as the electrons in the *n* material. The direct biasing voltage, however, never falls to zero, as at a certain point, the "tunnel" current is joined by the normal current produced by the flow of charge carriers due to the effect of the concentration gradients and the electric fields present within the junction. On the other hand, in the presence of an inverse biasing voltage (i.e. when the positive terminal is attached to the *n* region), the Esaki tunnel current i_e rapidly falls to zero, whereas the Zener current i_z continually increases. Therefore, considering the respective contributions of these two types of current, the observed characteristics of the tunnel diode can be explained. Figure 2.3 shows i_e, i_z and the total current i_{tot} as a function of the applied voltage.

Let us take a simple electronic circuit consisting of an ideal current generator and a tunnel diode. As the graphical analysis in Fig. 2.4 shows, within a certain range of currents this circuit presents three possible stationary states of which, as we shall see later, two are stable and the third unstable.

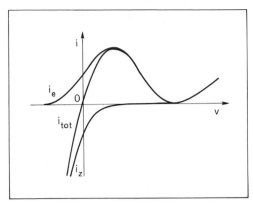

Fig. 2.3 - Esaki current i_e, Zener current i_z and total current i_{tot} versus applied voltage in a tunnel diode.

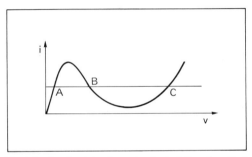

Fig. 2.4 - Steady states for a circuit consisting of an ideal current generator and a tunnel diode.

The static behaviour of the diode in a steady state condition is equivalent to that of a simple resistance equal to the ratio between the current and the voltage across the diode itself. In order to study the dynamic behaviour, however, it is necessary to consider not only the resistive but also the capacitive contribution. In fact, when the voltage across the diode is varied, there is a change in the charge stored in the space-charge region near the junction, the capacitance of which, in its turn, depends upon the applied voltage.

As this capacitance, over the range considered, depends to a lesser extent upon the voltage than does the resistance of the diode, we can simplify the problem by assuming that the value of the capacitance is constant.

The behaviour of the circuit can thus be described by a single non-linear differential equation. With C as the capacitance of the diode, i_t the current passing through it and v_t the applied voltage, the equation can be written as:

$$C \, dv_t/dt + i_t(v_t) = i_s \qquad (2.44)$$

where i_s is the current drawn from the generator.

The physical significance of this equation is quite simple: part of the current supplied by the generator flows through the diode (depending upon the applied voltage v_t) and the other part modifies the charge stored according to its capacitance C. Putting this stored charge equal to q_j

and $i(q_j) = i_s - i_t$, we can rewrite the above equation as:

$$dq_j/dt = i(q_j) \tag{2.45}$$

From an electronic point of view, this current can be divided into a charging current, $i_c(q_j)$, and a discharging current, $i_d(q_j)$. The equation can then be written as:

$$dq_j/dt = i_c - i_d \tag{2.46}$$

The charging current i_c (which we define as the current which tends to charge positively the p region with respect to the n region) is made up not only of the current from the external generator i_s, but also of the Zener current i_z which, as we know, corresponds to the flow of electrons from the valence band of the p region to the conduction band of the n region. Thus, we have: $i_c = i_s + |i_z|$.

In an analogous way the discharging current i_d consists of the Esaki current, i_e, which corresponds to the flow of electrons from the conduction band of the n region to the valence band of the p region, together with the normal direct current i_g of the p-n junction. In this case we have: $i_d = i_e + i_g$.

All of these currents, with the exception of i_s, are clearly functions of the voltage applied across the diode and, therefore, also functions of the charge stored in its capacitance. The differential equation which describes the behaviour of the circuit under consideration allows us to study the stability of the points of intersection between the diode characteristic and the load line. Given that v_0 is the voltage corresponding to a point of equilibrium, it is sufficient to evaluate the time evolution of a small change, Δv, that initially takes the system away from the value of v_0.

With $v_t = v_0 + \Delta v$ and developing i_t using first-order terms in Δv, we have:

$$C \, d \, \Delta v/dt + i_t(v_0) + (di_t/dv)_{v=v_0} \Delta v = i_s$$

and, since $i_t(v_0) = i_s$:

$$d \, \Delta v/dt = - \Delta v/\tau_0 \ , \tag{2.47}$$

where

$$\tau_0 = (di_t/dv)_{v=v_0} /C$$

The physical significance of $(di_t/dv) |_{v=v_0}$ is clearly that of an admittance for small signals about the value of v_0. One can see from Fig. 2.4 that $\tau_A > 0$, $\tau_B < 0$ and $\tau_C > 0$, so that points A and C are stable, whereas point B is unstable. This permits us to follow the dynamics of the circuit when approaching or leaving the points of equilibrium over a fairly close region around them.

The overall dynamics of the circuit, however, requires the solution of the nonlinear equation, which is necessary, especially for studying the "jump" from one state to another caused by a suitable external control signal.

There are, however, certain problems which this type of treatment is not capable of answering, particularly when the behaviour of the circuit depends upon the noise present within the circuit. An important problem, regarding the reliability of the tunnel diode as a two-state memory element, is the possibility of a spontaneous transition from state A to state C or vice versa.

Obviously, it is impossible to answer such questions using classical circuit equations due to the fact that, as already stressed in the previous discussion of the Nyquist theorem, these equations refer to the average values of macroscopic "order parameters" which refer to the collective behaviour of large numbers of electrons. Using the normal terminology for Brownian motion, in fact, we can say that electronic circuit equations only incl" de the "dissipation" effect of the innumerable interactions occurring without taking into account the simultaneous "fluctuation" effect which is inevitably linked to the former.

As a general result of our previous reasoning, we know that, in order to answer the above questions, the equations for the average values of the macroscopic order parameters must be transformed into Langevin equations with the introduction of "stochastic forces". This is in order to take into account the effect, upon the order parameters, of what we can call the "thermal bath" in which they are immersed.

Before introducing a Langevin equation for our order parameter (the charge q_j), it is worthwhile reformulating the problem, subjecting it to a collective type of treatment by means of a "master equation".

Such a treatment has an important physical significance, as we are dealing with a typical co-operative phenomenon. Tunnelling-induced transitions, in fact, depend upon the voltage applied across the diode; the single electrons, while not interacting with each other in a strict sense, do show an interdependent behaviour which is governed by a macroscopic order parameter.

In order to write the master equation, let us imagine a "set" of identical tunnel diode circuits of the type considered above. Let $n(N)$ be the number of members in the set in which the capacitive charge consists of N electrons. The variation of $n(N)$ with time depends upon the fact that there is a flow of electrons from one side of the device to the other, either through the device itself or through the external current generator. Thus we can consider the problem from a corpuscular point of view, taking time intervals so small that the possibility of two or more transitions taking place simultaneously can be ignored. With $\Gamma(N, N')$ as the transition rate from a state with N electrons to a state with N' electrons, and taking a small enough time interval, we can put:

$$[n(N, t + dt) - n(N, t)]/dt \simeq \partial n(N)/\partial t$$

We then have:

$$\partial n(N)/\partial t = \Gamma(N+1, N)\, n(N+1) + \Gamma(N-1, N)\, n(N-1) +$$
$$- \Gamma(N, N+1)\, n(N) - \Gamma(N, N-1)\, n(N) \tag{2.48}$$

For simplicity's sake, the time dependence of n has not been explicitly indicated. The physical significance of the transition rate is clearly the following:

$$\Gamma(N+1, N) = i_d\,(N+1)/q\,, \qquad \Gamma(N-1, N) = i_c\,(N-1)/q$$

In fact, the currents depend upon the charge stored by the capacitance, which is simply the product of the number of electrons and the electronic charge q. The fact that the transition involves only one electron at a time depends, as previously stated, upon the small time intervals considered.

The master equation can thus be rewritten as:

$$q\,\partial n(N)/\partial t = n(N+1)\, i_d\,(N+1) +$$
$$+ n(N-1)\, i_c\,(N-1) - n(N)\, i_d\,(N) - n(N)\, i_c\,(N) \tag{2.49}$$

Let us now suppose that the second member quantities are slowly varying functions of N, which is reasonable if N is large enough. It thus follows that, in the preceding equation, terms of the type $n(N+1)\, i_d(N+1) - n(N)\, i_d(N)$, which we will generally write as $\varphi(N+1) - \varphi(N)$, can be developed in Taylor series up to and including the second term:

$$\varphi(N+1) - \varphi(N) \simeq \partial\varphi/\partial N + (1/2)\, \partial^2\varphi/\partial N^2 + \dots$$

We then have:

$$n(N+1)\, i_d\,(N+1) - n(N)\, i_d\,(N) \simeq \partial[n(N)\, i_d\,(N)]/\partial N +$$
$$+ (1/2)\, \partial^2\,[n(N)\, i_d\,(N)]/\partial N^2$$

and also:

$$n(N-1)\, i_c\,(N-1) - n(N)\, i_c\,(N) \simeq \partial[n(N)\, i_c\,(N)]/\partial N +$$
$$+ (1/2)\,\partial^2\,[n\,(N)\, i_c\,(N)]/\partial N^2$$

Putting $N = x/q$ and introducing the ensemble density:

$$f(x,\ t) = n(x,\ t)/N_{tot}$$

we obtain the Fokker-Planck equation:

$$\partial f(x,\ t)/\partial t = -\,\partial[(i_c - i_d)f(x,\ t)]/\partial x + (q/2)\,\partial^2\,[(i_c + i_d)\,f(x,\ t)]/\partial x^2 \qquad (2.50)$$

We can also see that the preceding hypothesis gives us, in principle, a method for determining the limits of validity for the derivation of the Fokker-Planck equation from the master equation. The Fokker-Planck equation can also be obtained by starting from the Langevin equation. The latter equation can, in its turn, be obtained by simply adding a "stochastic force" term to the deterministic equation (2.46) for the "order parameter" q_j obtained above from elementary considerations.

The Langevin equation can thus be written as:

$$dq_j/dt = i_c - i_d + \phi \qquad (2.51)$$

where, as previously, i_c is the charging current and i_d is the discharging current.

The term $(i_c - i_d)$ thus represents a non-linear "force", because the currents depend upon the voltage across the device.

The term ϕ represents the fluctuations in the currents i_c and i_d (which are considered as mean values) and is not an ordinary function. However, with q as the electronic charge and supposing that the fluctuations in both directions are of equal intensity, we can, for example, use the following expression for the term ϕ:

$$\phi = q \sum_j (-1)^{n_j}\,\delta(t - t_j)$$

where n_j is a random variable taking the values of 0 or 1 with equal probability. If we now consider, as above, a set of analogous systems in which the probability density as a function of q_j is expressed by $f(q_j,\ t)$, we can write a Fokker-Planck equation (putting $q_j = x$ for simplicity):

$$\partial f(x,\ t)/\partial t = -\,\partial[K(x)f]/\partial x + (1/2)\,\partial^2\,[Q(x)f]/\partial x^2 \qquad (2.52)$$

where:

$$K(x) = \lim_{\Delta t \to 0} \langle x(t + \Delta t) - x(t)\rangle/\Delta t$$

$$Q(x) = \lim_{\Delta t \to 0} \langle[x(t + \Delta t) - x(t)]^2\rangle/\Delta t$$

In these expressions the symbol $\langle\,\rangle$, as usual, refers to an ensemble average.

Integrating the Langevin equation over a time interval which is small with respect to the characteristic time of x, but quite large with respect to the typical time of ϕ, we have:

$$x(t + \Delta t) - x(t) = \Delta x = (i_c - i_d)\Delta t + q \sum_{j=1}^{j_{max}(\Delta t)} (-1)^{n_j}$$

from which:

$$\langle\Delta x\rangle = (i_c - i_d)\,\Delta t \qquad (2.53)$$

since the ensemble average of the stochastic term is zero. If we then go on to evaluate $(\Delta x)^2$, we obtain:

$$(\Delta x)^2 = (i_c - i_d)^2 (\Delta t)^2 + 2q(i_c - i_d) \sum_{j}^{j_{max}(\Delta t)} (-1)^{n_j} \Delta t +$$

$$+ q^2 \left[\sum_{j=1}^{j_{max}(\Delta t)} (-1)^{n_j} \right]^2 \qquad (2.54)$$

The first two addends do not contribute to $Q(x)$; the former because it is of the second order in Δt, and the latter because it has a zero ensemble average. Let us now examine the third addend. The summation can take positive or negative values, but its square is always positive and therefore has a non-zero ensemble average. More exactly, as can be seen by imagining the development of the squared value followed by averaging, the ensemble average of such an expression over a time interval Δt is equal to the product of Δt and the average number of "granular events" (whether positive or negative) which take place over a period of time Δt. The ensemble average of the third addend can then be written as:

$$q^2 \Delta t (i_c + i_d)/q$$

where the currents are added instead of being subtracted from one another as in Eq. (2.53). On substituting in Eq. (2.53) we recover the form (2.50) for the Fokker-Planck equation. Thus $(i_c - i_d)$ can be interpreted as a drift term and $(q/2)(i_c + i_d)$ as a "diffusion coefficient". It is particularly significant, from the physical point of view, that in the "diffusion coefficient" the two currents add; this means that the fluctuations in i_c and i_d are independent of each other.

Let us briefly examine the steady state solution to the Fokker-Planck equation which now we rewrite as:

$$\partial f/\partial t = -\partial j/\partial x \qquad (2.55)$$

where:

$$j = (i_c - i_d)f - (q/2) \partial[(i_c + i_d)f]/\partial x \qquad (2.56)$$

Under stationary conditions we have $\partial f/\partial t = 0$, and thus $\partial j/\partial x = 0$, which is satisfied, in particular, when $j = 0$. On the other hand, the Fokker-Planck equation can be considered as a continuity equation, and the quantity j as an "ensemble current density" which clearly must be equal to zero under stationary conditions. The equation $j = 0$ (given that $i_c - i_d = \langle \dot{x} \rangle$ and that $(q/2) \cdot (i_c + i_d) = D$) can be written as:

$$f \langle \dot{x} \rangle - f \, dD/dx - D \, df/dx = 0 \qquad (2.57)$$

or as:

$$f V_x - D \, df/dx = 0 \qquad (2.58)$$

with

$$V_x = \langle \dot{x} \rangle - dD/dx \qquad (2.59)$$

We then have the expression

$$f = f_0 \exp \left[\int (V_x/D) \, dx \right]$$

in which D is function of x.

The current can no longer be written as the sum of a drift term and a diffusion term, which does not allow the verification of the fluctuation-dissipation theorem. As stated above, by means of the function f it is possible to study the global stability of the system. In the case of the tunnel diode, the form of this function is fairly complicated, even though it clearly can only take the form of a bimodal distribution. The "force" appearing in the Langevin and Fokker-Planck equations may be thought of as deriving from a "double-well" potential whose minima correspond to points A and C of Fig. 2.4.

2.4 COAGULATION IN COLLOIDAL SOLUTIONS

The phenomenon of coagulation in colloidal solutions is normally initiated by the addition of electrolytes and consists of the progressive aggregation of the suspension particles, with the consequent formation of aggregates of two, three, etc. single particles. The rank of an aggregate is defined as the number of elementary particles of which it is constituted.

The kinetics of this event was the object of Smoluchowski's study. He adopted a simplified model of the phenomenon which was essentially based upon three assumptions:

1) The condition necessary for aggregation is the approach of two particles (of whatever rank) until they reach a critical distance R. This distance may change according to the rank of the particles.
2) The above condition is also a sufficient one; in other words no distinction is made between "reactive" and "non-reactive" collisions. Such a distinction would imply, for example, the presence of energy barriers which could only be overcome by sufficiently violent collisions, or the existence of non-reactive states besides the reactive ones.
3) Only "bimolecular reactions" are allowed. That is, collisions involving more than two particles are ignored.

Given that every approach reaching the critical distance R is reactive, it is clear that the kinetics of coagulation will depend entirely upon the diffusion kinetics of the particles towards each other, and therefore the heart of the problem becomes the study of the diffusion process.

Colloidal particles are typical mesoscopic objects (in this sense they are exactly equivalent to Brown's pollen grains) and as such are sensitive detectors of fluctuations. Their motion is, therefore, Brownian and any analysis must be carried out in a probabilistic and not in a deterministic manner.

The problem is then formulated as follows: to determine the probability of finding a distance R between particles describing Brownian motion and therefore of predicting the time evolution of the particle population, or rather the probability of having n particles of rank i at time t.

From our treatment of free Brownian motion we know that the calculation of the probability density function for a single particle, $p(\vec{r}, \vec{v}, t)$, can be carried out by solving a Fokker-Planck differential equation. If we assume, however, conditions of sufficiently high viscosity, the velocities rapidly relax towards the equilibrium distribution so that it is sufficient to consider the time evolution of the particle position. The time evolution of the latter is much slower and can be described by a probability function $p(\vec{r}, t)$. The Fokker-Planck equation restricted to configuration space, also known as the Smoluchowski equation, takes the following form:

$$\partial p(\vec{r}, t)/\partial t = D\nabla^2 p(\vec{r}, t) \tag{2.60}$$

where D is the diffusion coefficient.

It should be noted that the greater simplicity of the Smoluchowski equation imposes some restrictions upon the variety of the possible boundary conditions and, consequently, upon the physical situations describable within this framework. The Smoluchowski equation, unlike the Fokker-Planck equation, does not allow the introduction into the model of restrictions on the velocity. For example, it is not possible to allow the absorption of a Brownian particle by a surface to depend upon the energy of the particle itself.

In the following Sections we will examine some physico-chemical problems in terms of the Smoluchowski equation. One of these problems (that of Brownian motion in the presence of absorbing boundaries) will be reconsidered in Section 2.7 under less restrictive conditions with the aid of the Fokker-Planck equation. This will give us the occasion to compare the results of both approaches and to discuss the nature and the extent of the discrepancies.

Eq. (2.60) is also known as a diffusion equation by structural analogy with the equation bearing the same name used to describe macroscopic diffusion processes:

$$\partial c(\vec{r}, t)/\partial t = D\nabla^2 c(\vec{r}, t) \tag{2.61}$$

where $c(\vec{r}, t)$ is the concentration of the chemical species which undergoes diffusion and D is the diffusion coefficient which indicates its mobility in the solution.

It is worth recalling that Eq. (2.61) results from two equations of a more elementary nature having an obvious physical significance. One is a continuity equation:

$$\partial c(\vec{r}, t)/\partial t = -\vec{\nabla} \cdot \vec{J} \tag{2.62}$$

and the other is known as Fick's law:

$$\vec{J} = -D\vec{\nabla} c(\vec{r}, t) \tag{2.63}$$

Eq. (2.62) determines that every change in concentration is inevitably accompanied by a flow of material $\vec{J}$ and is therefore transmitted to the surrounding regions. That is, matter can only be redistributed and neither created nor destroyed. Eq. (2.63) is a typical phenomenological law of the thermodynamics of irreversible processes which correlates the flow of material $\vec{J}$ with its distribution between adjacent points. That is, the spatial behaviour of the concentration function is locally represented by the concentration gradient and the flow $\vec{J}$ linearly depends upon this "spatial derivative"through the diffusion coefficient D. Given that

$$\vec{\nabla} \cdot \vec{\nabla} c(\vec{r}, t) = \nabla^2 c(\vec{r}, t)$$

Eq. (2.61) follows by substituting Eq. (2.63) for $\vec{J}$ in Eq. (2.62). The Smoluchowski equation for single Brownian motion can not, however be directly applied to our problem in which every collision is the result of two simultaneous Brownian motions. This difficulty can be overcome by the use of the following important theorem whose demonstration is omitted for the sake of brevity:

if two particles A and B moving independently of one another with Brownian motion have diffusion coefficients D_A and D_B, respectively, then the probability function for the relative position vector follows the equations of simple Brownian motion but with a diffusion coefficient $D = D_A + D_B$.

Therefore the basic elements of the Smoluchowski coagulation theory are Eq. (2.60) for single Brownian motion together with the theorem which states that two independent Brownian motions are equivalent to a single relative Brownian motion. As already mentioned Eq. (2.60) has a probability function as its argument and not a concentration as Eq. (2.61); we must therefore elucidate the connection between the non-measurable function $p(\vec{r}, t)$ and the concentration which is the fundamental observable in chemical systems.

Let us now examine only the collisions between two groups of particles of defined rank, say A and B. Following the theory mentioned above, we can consider each A particle as stationary with only the B particles describing Brownian motion. The function $p(\vec{r}, t)$ plays a role equivalent to that of a concentration if, instead of considering a particle B moving towards a particle A

(placed for simplicity at the origin of the axes), we consider a statistical ensemble of a large number of copies of the system composed of one particle A and N particles of type B, hereafter in short $(A + NB)$. The members of the statistical ensemble are chosen in such a way as to be representative of all the possible states of the system $(A + NB)$, in our case all the possible distributions of the B particles with respect to the A particle. Let us make some further assumptions in order to establish quite a simple relationship between the statistical ensemble and the real system. As far as the latter is concerned we shall consider only sufficiently dilute solutions so that mutual interactions among particles of whatever rank can be neglected.

Now the reactive interaction of each A particle will involve essentially the closest B particles (which we will call the "reactive cloud") while, in turn, the more remote B particles will have a very small probability of colliding with that particular A particle.

Moreover, assuming that the number of B particles is higher than that of the A particles so that each of the latter is surrounded by its own atmosphere of B particles, then it is reasonable to assume that the "reactive clouds" in the neighbourhood of distinct A particles are practically non-overlapping. Each B particle will then be assigned to only one "reactive cloud" at a time, and consequently competition among different A particles for the same B will be very unlikely.

The above hypotheses render the assumptions 1 to 3, stated at the beginning of this Section, more acceptable. On the basis of these assumptions the connection between the real system and the statistical ensemble becomes clear. In fact we can consider a colloidal solution, which is a collection of particles A and B, as a reasonable approximation of the statistical ensemble in the sense that it presents many of the possible positions of B with respect to A (clearly not all of them, but the most probable ones).

The statistical ensemble may be considered a contraction of the real system in the sense that all the A particles dispersed in the available volume become concentrated, in the ensemble, at the origin of the configuration space and therefore one studies the convergence of the various B particles towards a single super-particle A placed at the origin. In other words all the single collisions which take place at different points in the solution are referred to a single centre giving rise, in the statistical ensemble, to a flow of B particles towards the super-particle A. It is the magnitude of this flow towards the super-particle A which is measurable in the real system as the instantaneous rate of production of particles of higher rank. Therefore, $p(\vec{r}, t)$ gives the percentage of the ensemble systems wich have a B particle within the small volume $d\vec{r}$ centred at point $\vec{r}$.

From the way in which the statistical ensemble has been constructed, and assuming the problem demonstrates spherical symmetry, we can define the probability density per unit volume as:

$$p(r, t) = c(r, t)/c_A c_B \tag{2.64}$$

where $c(r, t)$ is the number of B particles in the real system which are to be found at a distance r from an A particle (or alternatively the concentration of B particles around the super-particle A), while c_A and c_B are the effective concentrations of A and B particles in the solution under equilibrium conditions.

The boundary conditions for Eq. (2.60) will be:

$$p(r, 0) = 1 \qquad \forall r > R \tag{2.65a}$$

$$p(r, t) = 0 \qquad r = R, \forall t \tag{2.65b}$$

where R is the critical distance for interaction between the two particles A and B.

Eq. (2.65a) reflects the additional hypothesis about the initial distribution of B particles which, accordingly, are assumed to be uniformly distributed throughout space at time $t = 0$. Non-uniform initial distributions can be handled in the framework of more refined mathematical treatments; however we will not dwell upon them as the very essence of the physical reasoning can be adequately illustrated in the simplest case of random initial distributions. Eq. (2.65b) expresses the fact that the A particles behave as ideal absorbers, in agreement with assumptions 1) and 3); in practice, this condition is approximately verified when the super-particle A has absorption rates much higher than the rate at which the B particles diffuse towards its surface. In this case the B particles do not collect on the surface of the super-particle.

Given that the cloud of B particles globally diffuses according to Eq. (2.60), the flow across the spherical surface of radius R around the super-particle A will follow Fick's law, Eq. (2.63), written for the function $p(r, t)$. The rate of association of the A particles with the B particles, equal to this flow, will be:

$$\Phi_{AB} = 4\pi R^2 D c_A c_B \, \partial p(r, t)/\partial r \, |_R \tag{2.66}$$

The solution of Eq. (2.60) subject to the conditions of Eqs. (2.65) is:

$$p(r, t) = 1 - \frac{R}{r} + \frac{2R}{r\sqrt{\pi}} \int_0^{(r-R)/2\sqrt{Dt}} \exp(-x^2)\,dx \tag{2.67}$$

by substituting Eq. (2.67) in the r.h.s. of Eq. (2.66) we obtain an explicit expression for Φ_{AB}:

$$\Phi_{AB} = 4\pi D R \, c_A c_B (1 + R/\sqrt{\pi Dt}) \tag{2.68}$$

In the above discussion we have limited ourselves by taking into account only two "species" of particles, A and B. Generally, we would have particles of rank i present in the suspension with a concentration c_i and every collision between a particle of rank i with one of rank j would give rise to a new particle of rank $(i + j)$ with unit probability. Thus, generalising Eq. (2.68), we reach the conclusion that the "reaction" rate between particles of rank i and rank j is:

$$\Phi_{ij} = 4\pi D_{ij} R_{ij} c_i c_j (1 + R_{ij}/\sqrt{\pi D_{ij} t}) \tag{2.69}$$

where R_{ij} is the critical distance between two particles of rank i and j.

This mechanism of the consumption of lower rank particles and the production of higher rank particles is closely analogous to a series of chemical reactions in which A_i particles of rank $i = 1, 2, \ldots$ take part as distinct chemical species, reacting according to the general scheme:

$$A_i + A_j \rightarrow A_{i+j} \qquad \forall \, i, j \tag{2.70}$$

For such reactions we can utilise the usual kinetic equations for the variation with time of the concentrations c_i. The rate constant for Eq. (2.70) can be deduced from Eq. (2.69) and is equal to:

$$k_{ij} = 4\pi D_{ij} R_{ij} (1 + R_{ij}/\sqrt{\pi D_{ij} t}) \tag{2.71}$$

and the kinetic equations will take the form

$$dc_i/dt = \frac{1}{2} \sum_{m+n=i} c_m c_n k_{mn} - c_i \sum_{j=1}^{\infty} c_j k_{ij} \tag{2.72}$$

Eqs. (2.72) comprise positive terms (the first summation), which represent the formation of particles of rank i from all possible collisions between particles of rank m and n such that $m + n = i$, and negative terms which take into account all the "reactions" in which particles of rank i take part as "reactants".

The solution of Eqs. (2.72) has been obtained by making the following approximations. First the time-dependent term in Eq. (2.71) is ignored so that

$$k_{ij} \simeq 4\pi R_{ij} D_{ij}$$

(under normal experimental conditions this condition is met after time lapses of the order of $10^{-3} - 10^{-4} s$ since the process started).

Furthermore on the basis of the observation that the diffusion coefficient, according to the Stokes equation, is inversely proportional to the radius of the particles and that the distance R is proportional to the radius associated with a single elementary particle, we can put

$$D_i R_i = D_1 R_1 \qquad (i = 1, 2, \dots);$$

if we also assume that

$$R_{ik} = (R_i + R_k)/2$$

and that

$$R_i = R_k$$

it can be deduced that

$$D_{ij} R_{ij} = 2DR$$

where $D = D_1$ and $R = R_1$

Including a change in the time scale

$$\dot{\tau} = (4\pi D R) t = \alpha t$$

with

$$u = 4\pi D R$$

Eqs. (2.72) can then be rewritten as

$$dc_k/d\tau = \sum_{i+j=k} c_i c_k - 2c_k \sum_{j=1}^{\infty} c_j \qquad (k = 1, 2, \dots) \tag{2.73}$$

From Eq. (2.73) we immediately obtain

$$\frac{d}{d\tau} \left(\sum_{k=1}^{\infty} c_k(\tau) \right) = - \left(\sum_{k=1}^{\infty} c_k(\tau) \right)^2$$

from which

$$\sum_{k=1}^{\infty} c_k(\tau) = c_0/(1 + c_0 \tau) \tag{2.74}$$

where

$$\sum_{k=1}^{\infty} c_k(0) = c_0 .$$

By using the integral in Eq. (2.74) we can successively integrate Eqs. (2.73) obtaining:

$$c_k = c_0 [(c_0 \tau)^{k-1}/(1 + c_0 \tau)^{k+1}] \tag{2.75}$$

The trends of the relative populations of the particles, c_k/c_0 against $c_0 \tau$ are shown in Fig. 2.5 for $k = 0, 1, 2, 3,$ and 4. These theoretical predictions turn out to be in good agreement with the experimental data.

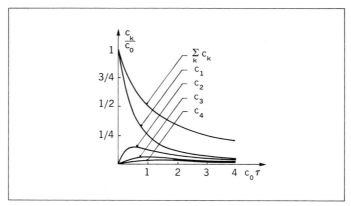

**Fig. 2.5 - Time evolution of the populations of colloidal particles of different rank
in the course of a typical coagulation process.**

2.5 DIFFUSION-CONTROLLED BIMOLECULAR REACTIONS

The rate of combination (Eq. (2.66)) is exclusively determined by the diffusion, in the sense that the particles are considered as totally inert and that they undergo an instantaneous aggregation as soon as they reach the critical distance R.

In chemical reactions, however, the reciprocal approach of the reagents is only the initial phase of the complete process which includes a chemical step (the actual chemical reaction) which involves the internal degrees of freedom of the molecules.

The reaction rates measured in a chemically active system thus include at least two contributions, a diffusive contribution and a chemical one, which, however, do not appear separately in the usual kinetic equations.

Certainly, there is no lack of examples of chemical reactions in which one of the two kinetics is negligible with respect to the other. In particular when the actual reaction under consideration is much faster than the diffusion process, then one can use the coagulation model and affirm that the measured reaction rate essentially reflects the diffusion step. If, however, the diffusion is fast with respect to the reaction time scale then it will be the latter which will essentially determine the experimental reaction rate.

Reactions having comparable times, which can no longer be clearly separated, for these two processes are called diffusion-controlled reactions. The theory of the reaction rate for this class of chemical events, which we will now examine in some detail for bimolecular reactions, was born as a re-elaboration of the Smoluchowski coagulation theory, and aims at distinguishing in the apparent rate, the diffusive contribution from the effective chemical rate.

One way of reproducing the effect of the complex interactions which determine the reaction rate consists of the assumption that only a fraction of the collisions are effective. If the Smoluchowski model is to be adapted to the treatment of bimolecular reactions, it is clearly necessary to abandon assumption 2) and to modify Eq. (2.66) by introducing a factor γ, with $0 < \gamma < 1$, which represents the fraction of useful collisions.

Including γ is equivalent to accepting that the absorber is no longer ideal but partially reflective. There then arises an obvious contradiction with the condition expressed by Eq. (2.65b) which asserts that no particles rest upon the surface of the absorber. This inconsistency can be eliminated by substituting Eqs. (2.65) with a new boundary condition which takes into account the influence of the "penetration" rate (which is no longer infinite) of the B particles into the super-particle A upon the diffusive flow (Eq. (2.66)).

Thus it is required that:

$$\Phi_{AB} = kp(R,t)\, c_A c_B \qquad (2.76)$$

Eq. (2.76) essentially represents a flow entering the super-particle A, which, according to the rules of chemical kinetics, is proportional to the probability of finding B at a distance R from the super-particle. The constant of proportionality, K, is inherent in the chemical reaction mechanism.

Combining Eq. (2.76) with Eq. (2.66) gives the new boundary condition

$$K\,p(R,\ T) = 4\pi R^2 D\ \partial p(r,\ t)/\partial r\ |_R \qquad (2.77)$$

Eq. (2.60) with this new boundary condition can be resolved and it is interesting, without analysing the solution in detail, to reflect upon the limiting cases for $K \to 0$ and $K \to \infty$.

$K \to 0$ means that the chemical kinetics are much slower than the diffusion kinetics so that the diffusion phenomena, much faster than the chemical ones, have no bearing upon the reaction rate and effectively maintain the osmotic equilibrium so that the flow in Eq. (2.76) reduces to:

$$\Phi_{AB} = Kp(R,\ 0)\, c_A c_B = K c_A c_B$$

in agreement with deterministic kinetics (under equilibrium conditions $p(R,\ t) = p(R,\ 0) = 1$).

On the other hand, with $K \to \infty$ the chemical kinetics become very fast with respect to the diffusion kinetics and the results converge with those of Smoluchowski, which were obtained by considering only the diffusion as the kinetically dominating phenomenon. In this case the reaction is said to be completely diffusion-controlled.

The theory of diffusion-controlled reactions is applied here to the study of a specific case of biochemical interest, represented by a wide class of reactions between macromolecules and ligands which can freely diffuse in solution. In particular, we will consider the reaction between a protein (myoglobin or haemoglobin) and a ligand (oxygen).

In order to simplify the calculation the protein will be considered as an isotropic sphere characterised by a uniform reactivity over all its surface. This approximation, which clearly detracts from the well-known anisotropy of proteins in general can, however, be substituted by a more realistic model in which only a limited sector of the sphere is reactive. We will develop here the isotropic model because from this model one can obtain, in quite a direct manner, results which can then be extended to the anisotropic case.

The Smoluchowski equation, in spherical coordinates, is:

$$\partial(rp(r,\ t))/\partial t = D\ \partial^2(r\,p(r,\ t))/\partial r^2 \qquad (2.78)$$

where D includes the diffusion coefficients both of the ligand and of the protein.

If R is the critical distance between ligand and protein, the boundary conditions for Eq. (2.78) are still given by Eq. (2.66a) and Eq. (2.77):

$$p(r,\ 0) = 1 \qquad\qquad \forall r > R \qquad (2.79a)$$

$$Kp(R,\ t) = 4\pi R^2 D\ \partial p(r,\ t)/\partial r\ |_R \qquad (2.79b)$$

which we complete with

$$p(\infty,\ t) = 1 \qquad\qquad \forall t \qquad (2.79c)$$

which describes the continuation of the state of equilibrium at infinity even during the reaction (chemical and diffusion processes significantly alter the equilibrium distribution of the ligand only at finite distances). The time-dependent association coefficient, according to Eq. (2.66), is

$$K_1(t) = 4\pi R^2 D\ \partial p(r,\ t)/\partial r\ |_R \qquad (2.80)$$

where $p(r, t)$ is the solution to Eq. (2.78) under the conditions of Eqs. (2.79). This solution is given by

$$p(r, t) = 1 - \frac{R-\beta}{r} \left\{ erfc\left[\frac{r-R}{\sqrt{4Dt}}\right] + \exp\left[\frac{Dt}{\beta^2} + \frac{r-R}{\beta}\right] erfc\left[\frac{\sqrt{Dt}}{\beta} + \frac{r-R}{\sqrt{4Dt}}\right] \right\}$$

where

$$1/\beta = (K/4\pi R^2 D) + (1/R)$$

and

$$erfc\, x = \frac{2}{\sqrt{\pi}} \int_x^\infty \exp(-z^2)\, dz$$

For the following discussion it is sufficient to use the asymptotic value of K_1

$$K_\infty = \lim_{t \to \infty} K_1(t) = KK_D(K_D + K)^{-1} \tag{2.81}$$

with

$$K_D = 4\pi R D. \tag{2.82}$$

2.6 REACTIVITY MODULATION IN DIFFUSION-CONTROLLED BIMOLECULAR REACTIONS

For bimolecular reactions of the type

$$A + B \xrightarrow{K}$$

governed by the kinetic equation

$$dc_A/dt = -K_{tot}\, c_A c_B = dc_B/dt$$

the theory of diffusion-controlled reactions allows the separation of the intrinsic reaction constant K from the influence of the diffusion process occurring before the reaction itself which are both included in the global reaction constant K_{tot}, which is directly measurable.

The distinction of these two successive steps may be insufficient in accounting for the experimental data. This is the case for the reaction between myoglobin (haemoglobin) and oxygen in which a discrepancy of about one order of magnitude (lower) was found between the observed rate and that which would be expected were the reaction completely diffusion-controlled. This discovery suggests the intervention of a new step, intermediate between the diffusion step and the reaction step, which slows down the process to the observed levels. It has been conjectured that the new step consists in a deformation of the macromolecule which involves an oscillation of the lateral chains of the protein alternately covering and then uncovering a kind of access route to the active site.

We shall illustrate a deterministic model for this modulation of the reactivity, in which the macromolecule oscillates between a reactive state and an inert state, with a rhythm specified by a time-dependent function $h(t)$.

In this way the access to the active site of the protein is considered as alternately open ($h = 1$) or closed ($h = 0$); furthermore if we ignore the kinetic details of the transition from the "open" state O to the "closed" state C, i.e. if the transition is assumed to be almost instantaneous, then $h(t)$ takes the form of a square wave.

The presence of the modulation affects the boundary condition of Eq. (2.79b) which then becomes

$$K p(R, t) h(t) = 4\pi R^2 D\, \partial p(r, t)/\partial r \,|_R \qquad (2.83)$$

With $h = 1$ Eq. (2.83) corresponds to Eq. (2.79b) whereas with $h = 0$ Eq. (2.83) implies that

$$\partial p(r, t)/\partial r \,|_R = 0 \qquad (2.84)$$

which is the appropriate boundary condition for Brownian motion in the presence of a perfectly reflecting surface at $r = R$ (Eq. (2.66b)). Once $h(t)$ is known, the average entrance flow, that is, the reaction rate, can be calculated, at least in principle. If, however, it is assumed that the periods of "closure" are sufficiently long so that at every "re-opening" the initial conditions of equilibrium are recovered, $(p(r, t) = 1$, see Eq. (2.79a)), then the solution to the modulated problem can be taken back to the unmodulated case.

If the site is "open" for an arbitrary time t_O and "closed" for a period which is also arbitrary but sufficiently long, t_C (where t_O and t_C are constant), then the modulated rate constant is the time average of the flow entering the protein, and that is:

$$K_m = (t_A + t_C)^{-1} \int_0^{t_O} K_1(t)\, dt = \langle h \rangle K(t_O) \qquad (2.85)$$

where

$$\langle h \rangle = t_O/(t_O + t_C)$$

is the average period of time that the protein remains in state O and

$$K(t_O) = t_O^{-1} \int_0^{t_O} K_1(t)\, dt$$

is the average flow during the "open" period ($K_1(t)$ is the rate constant in the absence of modulation, see Eq. (2.80)).

The calculation can be directly extended to the case where the "open" periods are not always equal to t_O but are distributed in a random manner according to a known probability density function $\varphi(t_O)$. Naturally the constraint

$$t_O \ll t_C$$

must be maintained.

A more refined modelling of the same system can be given where the restriction on the length of the "closure" periods is no longer necessary. In this version, undoubtedly more realistic on a molecular scale, the modulation mechanism is a stochastic one. This means that the molecule oscillates between states O and C in a way that can no longer be described by a time-dependent function but only in terms of probabilities. We can visualize the change of state of the protein as a kind of reversible reaction

$$O \underset{b}{\overset{a}{\rightleftarrows}} C \qquad (2.86)$$

The reaction constants a and b are transition probability functions per unit time which specify, respectively, the probability that the protein undergoes the transition $O \to C$ or $C \to O$ with the protein initially in state O or C.

The stochastic process which regulates the transitions of Eq. (2.86) is also assumed to be Markovian and stationary. That is to say:

$$\partial a/\partial t = \partial b/\partial t = 0$$

If, in unit time, we have, on the average, a or b "closures" or "openings", then the average dwell time in state O or state C is $1/a$ or $1/b$ and the average length of one cycle

$$A \to C \to A$$

is consequently

$$1/a + 1/b = (a + b)/ab \qquad (2.87)$$

whereas $ab/(a + b)$ represents the average number of cycles per unit time. On average, the protein will be in state O for a fraction of the unit time, equal to the probability of the state itself,

$$ab/[a(a + b)] = b/(a + b) \qquad (2.88)$$

Also, the protein will be found in state C with a probability of

$$ab/[b(a + b)] = a/(a + b). \qquad (2.89)$$

Stochastic modulation introduces a second stochastic factor besides that already incorporated in the Smoluchowski equation (that is, the Brownian motion of the ligand towards the protein). The continuous and irregular oscillation of the protein between state O and C (gating) influences the boundary conditions for Eq. (2.78) as in the deterministic case (Eq. (2.83)); only now there is no function describing the temporal behaviour of the oscillation so that it is not possible to condense the random alternation of the boundary conditions, expressed by Eqs. (2.79b) and (2.84), into a single equation equivalent to Eq. (2.83). Summarising, the fluctuation of the protein between the two states O and C leads to the alternation of two distinct forms for the boundary conditions.

The problem of solving a differential equation with "fluctuating" boundary conditions may be converted into a normal problem with fixed boundary conditions at the cost of transferring, in a certain sense, the fluctuation of the boundary conditions to the function $p(r, t)$. This artefact consists of attributing the change of state (Eq. (2.86)) to the ligand, whereas in reality it relates to the protein. In practice, the configuration space of a single molecule of the ligand is enlarged so that every state of the ligand is then identified by the three spatial coordinates together with a new coordinate s which can take the "value" O and C. Correspondingly, the function $p(r, t)$ then depends upon the new variable s and thus becomes a $w(r, s, t)$ function.

For the $w(r, s, t)$ function the diffusion in the three-dimensional subspace (x, y, z) is still governed by Eq. (2.78) except that new terms are added to the equation in order to take into account the changes in $w(r, s, t)$ with respect to time related to the "diffusion" between the two subspaces corresponding to $s = O$ and $s = C$.

This flux of $w(r, s, t)$ in the s-subspace is regulated by the transition probability functions a and b; starting from them the most suitable formal instrument for describing the variations in the density w in phase space is, as we recall, a master equation.

Arriving step by step at the complete equation, we will assume, for the moment, that the kinetics of diffusion are much faster than the process represented by Eq. (2.86). We will consider a time scale in which the slower phenomenon is practically stationary and focus our analysis on the faster process.

We will then observe a sequence of reactions with diffusion when the protein is in state O alternating with simple diffusion when the protein is in state C. The changes of state, however, will be sufficiently infrequent to allow the stabilization of diffusion phenomena over the stationary states compatible with the instantaneous boundary conditions.

The equation which governs the evolution of the system will still be Eq. (2.78) with boundary conditions which vary with time, but so slowly that they can be considered as constant while the diffusion restores the stationary state or the state of equilibrium according to whether the protein is to be found in state O or state C.

If, on the other hand, the gating were much faster than the diffusion then the protein would undergo a large number of transitions (2.86) in such short times that the diffusion would not be able to significantly alter the spatial distribution of the ligand. The transition between states O and C would obey a master equation of the following type:

$$\partial p_O(r, t)/\partial t = -a p_O(r, t) + b p_C(r, t)$$
$$\partial p_C(r, t)/\partial t = a p_O(r, t) - b p_C(r, t)$$

(2.90)

From this point onwards, the coordinate s will be given as a subscript of the function $p(r, t)$ according to the convention

$$w(r, O, t) = p_O(r, t)$$
$$w(r, C, t) = p_C(r, t)$$

Eq. (2.90) immediately acquires a more familiar aspect when we return to the chemical metaphor and consider the process (2.86) as a reaction which converts a reagent O into a product C. Eq. (2.90) is really a pair of kinetic equations having the "concentrations" p_O and p_C as their arguments.

In conclusion, if we want to study a reaction where the diffusion and gating kinetics do not have separate time scales, but, on the contrary, clearly interfere with one another, we retain the terms deriving from the Smoluchowski equation and Eq. (2.90) obtaining the global equation:

$$\partial p_O(r, t)/\partial t = -a p_O(r, t) + b p_C(r, t) + \nabla^2 p_O(r, t) \cdot D$$
$$\partial p_C(r, t)/\partial t = a p_O(r, t) - b p_C(r, t) + \nabla^2 p_C(r, t) \cdot D$$

(2.91)

The structure of Eq. (2.91) is of a mixed type in the sense that for the diffusion process the Fokker–Planck approximation has been adopted whereas for the conformational fluctuation of the protein the original master equation has been retained.

The quantity in which we are interested is the stationary stochastic association constant which, by analogy with Eq. (2.80), is equal to

$$K_{Stoch} = 4\pi R^2 D \partial p_O(r, t)/\partial r |_R$$

(2.92)

where $p_O(r, t)$ is the solution to Eq. (2.91) subject to the boundary conditions:

$$\partial p_C(r, t)/\partial r \Big|_R = 0$$

(2.93a)

$$4\pi R^2 D \partial p_O(r, t)/\partial r |_R = K p_O(R, t)$$

(2.93b)

$$p_O(\infty, t) = b/(a + b) \qquad \forall t$$

(2.93c)

$$p_C(\infty, t) = a/(a + b) \qquad \forall t$$

(2.93d)

It is to be noted that the r.h.s. of Eqs. (2.93c) and (2.93d) coincide with Eqs. (2.88) and (2.89).

By then solving Eq. (2.91) under stationary conditions

$$\partial p_O(r, t)/\partial t = \partial p_C(r, t)/\partial t = 0$$

we obtain

$$p_O(r) = -\frac{X}{r} \exp\left\{-\sqrt{\frac{a+b}{D}}\, r\right\} + \frac{b\,Y}{r} + \frac{b}{a+b}$$

$$p_C(r) = \frac{X}{r} \exp\left\{-\sqrt{\frac{a+b}{D}}\, r\right\} + \frac{a\,Y}{r} + \frac{a}{a+b}$$

(2.94)

where X and Y are constants determined on the basis of the conditions expressed by Eqs. (2.93). Note that, in agreement with Eqs. (2.93c) and (2.93d)

$$\lim_{r \to \infty} p_O(r) = b/(a + b)$$

$$\lim_{r \to \infty} p_C(r) = a/(a + b).$$

Calculating the constants X and Y, and substituting the first Eq. (2.94) in Eq. (2.92) we obtain the stationary stochastic association constant:

$$K_{stoch} = \frac{K_D K b Z(a + b)}{a[K + K_D Z(a + b)] + b(K_D + K) Z(a + b)}$$ (2.95)

where the function $Z(s)$ is defined as

$$Z(s) = 1 + \sqrt{st_D}$$

where

$$t_D = R^2/D$$ (2.96)

is a characteristic time for the diffusion. Eq. (2.95) allows a verification of the validity of the intuitive considerations which led to Eq. (2.91) and which relate to the limiting cases where the reactivity modulation and diffusion occur on widely differing time scales.

Let us suppose that the times for the process represented by Eq. (2.86) are long with respect to the typical times for the diffusion t_D, (Eq. (2.96)), that is:

$$(a + b)^{-1} \gg t_D$$

Then $Z \to 1$ and Eq. (2.95) reduces to

$$K_{Stoch} = b(a + b)^{-1} K_\infty.$$

This means that the stationary stochastic constant coincides with the stationary constant for the unmodulated problem (see Eq. (2.81)) reduced by a factor equal to the probability that the protein is to be found in state O. This result is reasonable given that, on the average, over sufficiently long periods of time, the protein is reactive only for a fraction $b/(a + b)$ of the period under consideration. If, however, the reactivity modulation is fast with respect to the diffusion, and therefore

$$(a + b)^{-1} \ll t_D,$$

then $Z \to \infty$ and consequently

$$K_{Stoch} = K_D Kb[(a + b) K_D + bK]^{-1}$$ (2.97)

Let us remark that the magnitude of the intrinsic reaction constant K, in this situation of fast modulation, could have a sizeable effect on the overall behaviour of the reaction. More precisely, if $K \to \infty$ (completely diffusion-controlled reaction) then

$$\lim_{K \to \infty} K_{Stoch} = K_D$$

where K_D, given by Eq. (2.82), coincides exactly with the asymptotic value for $K \to \infty$ of the constant for the unmodulated problem (Eq. (2.81)). Finally, if the reaction is not diffusion-controlled, that is, if $K \to 0$ and therefore if $K \ll K_D$, then Eq. (2.97) becomes:

$$K_{Stoch} = K b (a + b)^{-1}$$

In this case only the intrinsic reaction constant and the gating kinetics contribute to the stochastic constant while the diffusion is fast enough to restore the equilibrium distribution over much shorter periods of time.

The previous treatment may be generalised for the study of stochastic problems having other than spherical geometries or for more realistic models which contemplate a reactivity limited to certain sectors of the total macromolecular surface.

2.7 BROWNIAN MOTION IN THE PRESENCE OF ABSORBING BOUNDARIES

It has been noted that many problems involve the absorption of Brownian particles by an ideal or semi-reflective absorber in which case prediction of the kinetics is reduced to an evaluation of the flow entering the absorber.

In many cases, however, it may be interesting to examine not only the time evolution of the process but also the spatial distribution of the particles around the absorber.

Evaluation of the concentration profiles is, to cite a concrete example, fundamental for membrane biophysics in general, and in particular for the calculation of membrane potentials or, in any case, for all of those situations in which concentration measurements carried out at large distances from the membrane can not supply reliable information on concentrations near the membrane precisely because of the strong concentration gradients which are formed in solution.

Given the importance of the matter we will study the case of Brownian particle absorption by a spherically symmetric absorber which, referring back to the discussion in Section 2.6, we will assume to be a perfect absorber. It is to be noted here that, as there is only one absorber, the probability density function will now be directly correlated with the actual particle concentration so that the Smoluchowski equation (Eq. (2.60)) and the diffusion equation (Eq. (2.61)) with the conditions of Eqs. (2.65) (or analogous conditions in terms of concentrations) are practically coincident.

The condition of Eq. (2.65b) (which from here onwards we shall call the Smoluchowski condition), was derived by Smoluchowski for a random walk in the presence of an absorbing boundary and defines the perfect absorber by attributing to it absorption kinetics infinitely faster than the diffusion kinetics.

Such a condition is obviously too drastic in many real examples where such a clear separation between the two kinetics does not exist. A more realistic boundary condition would be to put equal to zero only the concentration of the emergent particles and to keep non-zero the concentration of the entering particles.

Such a distinction may be made only in terms of the velocity variable, defining the two particle currents with the inequalities

$$v > 0 \quad \textit{emergent particles}$$
$$v < 0 \quad \textit{incident particles}$$

(2.98)

For simplicity, we will, for the moment, limit the discussion to the one-dimensional case, allowing the possibility of generalising the treatment to the three-dimensional problem later.

Now, the most obvious intrinsic limit of Eq. (2.60) is that it allows modelling limited only to configuration space thus precluding the possibility of carrying out velocity discriminations of the type in Eq. (2.98) or of a similar type which would allow the representation, for example, of a non-perfect absorbing boundary for which the absorption or the reflection of the particles could depend upon suitable threshold values of the velocities.

The most direct way of including information on the velocity variable consists in the adoption of a Fokker-Planck equation for the phase space of the single Brownian particle, in place of the Smoluchowski equation.

These considerations bring us back to the problem, widely discussed in the literature, of passing from the more detailed description provided by the Fokker-Planck equation to the more limited representation given by the Smoluchowski equation. One example of a solution is that of Kramers, already discussed in Section 1.6, while later work has contributed towards both finding further approximations and confirming, in a more rigorous manner, Kramers' result. Particular attention was paid to the description of the transient from the Fokker-Planck regime to the Smoluchowski (or diffusive) one (in terms of perturbative expansions which provide corrective terms for the Smoluchowski equation) as well as to the problem of finding suitable boundary conditions for both the former as well as the latter equations.

Without going into detail, it is worth recalling that some analyses of the time-dependent Fokker-Planck equation have examined the convergence of the solution to that of the related Smoluchowski equation showing the persistence of certain discrepancies between the two types of solution in the presence of various boundary conditions and using different initial conditions for the velocity. In particular, local concentration values may differ to a significant degree when the initial conditions for the velocity are of a non-Maxwellian type. These differences are particularly notable in the proximity of absorbing or reflecting boundaries where one expects large deviations from the Maxwellian velocity distribution. The Fokker-Planck equation in question may be written by starting from a two-dimensional stochastic Langevin equation for the process (x, v)

$$v = dx/dt$$
$$dv/dt + \beta v = F(t) \tag{2.99}$$

The Fokker-Planck equation for the joint probability density function has the same coefficients as Eq. (1.48) and actually follows from Eq. (1.48) integrated over the intial state (x_0, v_0). This gives:

$$\frac{\partial p}{\partial t}(x, v, t) = \left[-\frac{\partial}{\partial x} v + \beta \frac{\partial}{\partial v} v + D \frac{\partial^2}{\partial v^2} \right] p(x, v, t) \tag{2.100}$$

where D is the diffusion coefficient related to the autocorrelation function of the stochastic force $F(t)$ by the relationship

$$\langle F(t) F(t + \tau) \rangle = 2D\delta(\tau)$$

Considering only the stationary case, for which

$$\frac{\partial p(x, v, t)}{\partial t} = 0$$

Eq. (2.100) becomes:

$$v \frac{\partial p(x, v)}{\partial x} = \beta \frac{\partial}{\partial v} \left(\frac{D}{\beta} \frac{\partial}{\partial v} + v \right) p(x, v) \tag{2.101}$$

This increase in the set of independent variables offers the advantage of allowing definition of the boundary conditions regarding the actual velocity in an explicit manner and thus, in the case of a perfectly absorbing boundary at $x = 0$, we can use the boundary condition

$$p(0, v) = 0 \qquad \forall v > 0, \qquad \forall t \tag{2.102}$$

which has an obvious physical meaning.

This condition is of the type commonly referred to in the literature as "half-range" as it relates only to the positive velocities. The solution of Eq. (2.101) in this case presents serious difficulties precisely because of the "full-range" character of the equation and the "half-range" character of the boundary condition.

In order to arrive at solutions which satisfy the condition of Eq. (2.102) methods have been proposed which are based upon the use of complete sets of orthogonal functions defined over a half-space. Analyses inspired by different techniques, both of a numerical and an analytical nature, carried out on the Fokker-Planck equation, have been found to be very useful for a constructive criticism of the boundary conditions to be used for the Smoluchowski equation. The attempt to check the deducibility of the condition given by Eq. (2.65b) was unsuccessful: in fact the Fokker-Planck equation and the related condition (Eq. (2.102)) imply non-zero probability on the boundary. On the other hand it has been found that, by imposing this modified boundary condition to the Smoluchowski equation, the thickness of the boundary layer within which the discrepancies between the solutions of the Fokker-Planck and the Smoluchowski equations are more notable, actually decreases. Considerations of an analytical nature based upon an assumed structure of the solution suggest that in the diffusion limit Eq. (2.101) reduces to Eq. (2.77). Thus Eq. (2.77) has a more solid basis than the arguments of plausibility with which it was previously justified. This result, proven for a one-dimensional system, is also valid for spherical geometry and, also, the procedure adopted turns out to be equivalent to the "half-range" procedure which we will use below, in the sense that both give the same estimates for the first moments of the probability distribution.

For the "half-range" technique it is assumed that the function $p(x, v)$ has a different structure depending upon whether $v > 0$ or $v < 0$.

The basic hypotheses for the structure of the function $p(x, v)$ (also common to other methods which use, for example, expansions in full-range eigenfunctions of the Fokker-Planck operator), may be summarised as follows: the $p(x, v)$ function is approximately a Maxwellian distribution containing auxiliary functions n_i and u_i (which depend only upon the spatial coordinate) of the following type

$$p_i(x, v) = \frac{n_i(x)}{\sqrt{\pi\gamma}} \exp\left[-(v - u_i(x))^2/\gamma\right] \qquad (2.103)$$

$$i = 1 \quad \text{if} \quad v > 0$$

$$i = 2 \quad \text{if} \quad v < 0$$

$$\gamma = 2kT/m$$

The imperfectly Maxwellian form and the dependence of p_i upon the corrective parameters $n_i(x)$ and $u_i(x)$ are equivalent to the assumption of a non-equilibrium state which it is reasonable to expect, especially close to the absorbing surface. The surface at $x = 0$ absorbs particles but does not re-emit them into the environment so that, close to the surface, an atmosphere is formed of particles which are mainly entering (having negative velocities) and thus a state of non-equilibrium is formed (under conditions of equilibrium the two sets of velocities $v > 0$ and $v < 0$ would, in fact, be equally represented at each point in space).

The four functions $n_i(x)$ and $u_i(x)$ are determined by setting up the equations for the first four moments of $p(x, v)$.

The moments of the $p(x, v)$ function are defined as

$$m_n(x) = \int v^n \, p(x, v) \, dv$$

and their evolution equations are constructed by multiplying both sides of Eq. (2.101) by v^n and then integrating over the velocities. The resulting equations for the first four moments are:

$$dm_1/dx = 0$$

$$dm_2/dx + m_1\beta = 0$$

$$\frac{dm_3}{dx} + 2\beta m_2 - 2D\beta^2 m_0 = 0$$

$$\frac{dm_4}{dx} + 3\beta m_3 - 6D\beta^2 m_1 = 0$$

(2.104)

where

$$m_0(x) = c(x) = particle\ concentration$$

and

$$m_1(x) = j(x) = particle\ flow.$$

The initial conditions for Eq. (2.104) are:

$$n_1(0) = 0 \tag{2.105a}$$

$$j(L) = A \tag{2.105b}$$

$$\lim_{x \gg \ell} c(x) = Ax + B \tag{2.105c}$$

where

$$\frac{c(x)}{c_0} = p(x) = \int p(x,\ v)\ dv \tag{2.105d}$$

and c_0 is the equilibrium particle concentration. ℓ is a measure of the boundary layer thickness, i.e. the region where the concentration profile differs from the bulk profile by a factor proportional to exp $[-\ell x]$. ℓ turns out to be of the order of magnitude of the distance travelled by a Brownian particle with average velocity $\langle v \rangle \sim \sqrt{kT/m}$ during a time β^{-1} (equal to the velocity relaxation time): thus $\ell \sim \sqrt{kT/m}\ /\ \beta$.

Eq. (2.105a) ensures that the condition (2.102) is met and Eq. (2.105b) represents the entity of the flow which is sustained at the boundary of the system for $x = L$ in order to guarantee the stationary nature of the distribution, while (Eq. (2.105c) expresses the convergence of $c(x)$ with the function $Ax + B$ (which is the solution of Eq. (2.60)) for distances from the absorbing surface much greater than ℓ.

To aid the calculation Eq. (2.103) may be slightly modified in order to avoid equations which are extremely nonlinear and which can not be dealt with using normal techniques. Given that in general one has to operate under the conditions

$$|u_i(x)| \ll \sqrt{\gamma}$$

then in the exponential function of Eq. (2.103) exp$(-v^2/\gamma)$ can be extracted and, neglecting exp $(-u_i^2/\gamma)$, we can make the approximation

$$\exp\{[-u_i^2(x) + 2vu_i(x)]/\gamma\} \approx 1 + 2vu_i(x)/\gamma$$

Thus Eq. (2.103) becomes

$$p_i(x, v) = \frac{n_i(x)}{\sqrt{\pi\gamma}} \left[1 + \frac{2vu_i(x)}{\gamma} \right] \exp(-v^2/\gamma) \tag{2.106}$$

The preceding calculation may be adapted to the three-dimensional case with spherical symmetry in order to describe the absorption of Brownian particles by a perfect spherical absorber of radius R. Without reporting the detailed calculations which may be developed by analogy with the one-dimensional case, we focus upon the radial concentration profile which results to be:

$$c(r) = c_0 \left\{ 1 - \frac{R}{r} \ell R (1 + 1.85 \ell R) \frac{4}{3\pi C(R)} - \frac{4}{\pi} \frac{A(r)}{A(R) C(R)} \exp\left[-\ell(r-R)\right] \cdot \right. $$
$$\left. \cdot \left[\frac{R}{r} \ell R \, 0.35 + \frac{R^2}{r^2} \right] \right\} \tag{2.107}$$

with

$$A(r) = \left[1 + \frac{8\ell r}{3\pi} \right]^{(6\pi/15 - 1)}$$

and

$$C(R) = 0.79 \, (\ell R)^2 + 2.35 \, \ell R + 2.27$$

It is interesting at this point to compare Eq. (2.107) with the profile predicted by Eq. (2.60) which we obtain as the limit for $t \to \infty$ from Eq. (2.67), i.e.

$$c_{DE}(r) = c_0 (1 - R/r) \tag{2.108}$$

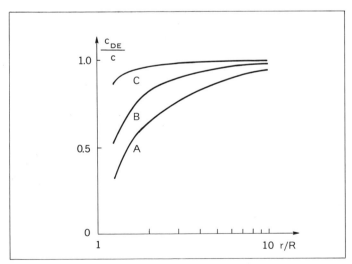

Fig. 2.6 - **Ratio of the concentration profiles calculated from the Smoluchowski equation (c_{DE}) and the Fokker-Planck equation (c). Curves A, B and C refer respectively to $\ell R = 2, 10, 100$ (after Harris, 1982).**

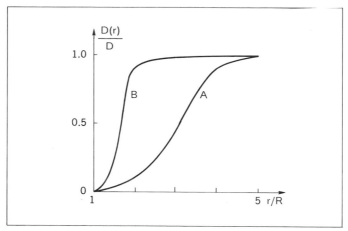

Fig. 2.7 - **Behaviour of the space-dependent diffusion coefficient** $D(r)$ **(normalized to the constant diffusion coefficient** D **of the Smoluchowski equation) for** $\ell R = 2, 10$ **(curves** A **and** B **respectively) (after Harris, 1982).**

The differences between Eq. (2.107) and (2.108) are illustrated in Figure 2.6. It can be seen that Eq. (2.108) leads to an underestimation of the concentration which is more notable in the boundary layer adjacent to the absorbing sphere and having a thickness of the order of $10R$.

High values of the product ℓR mitigate the thickening effect predicted by Eq. (2.107) which, for $r \to \infty$, converges towards the values of Eq. (2.108).

Finally, we recall that, close to the absorber, it is to be expected that correlations between the molecular motions of the solvent and the motion of the absorber become significant so that the constant coefficient D of the Smoluchowski equation needs to be replaced with a coefficient $D(r)$ which depends upon the radial coordinate.

Analytical evaluations of $D(r)$ have been made, being founded upon a continuous hydrodynamic model of the solvent, but the fluid can clearly not be modelled in this way when the volumes being considered are of the order of the molecular volume (consider the case where $R \ll 1$), whereas the microscopic calculations of $D(r)$ would be extremely complicated.

Computer simulations have been carried out using a discrete model of the solvent which provided numerical estimates of $D(r)$, whereas other attempts have used $D(r)$ functions chosen on the basis of their plausibility albeit in rather an arbitrary manner.

The calculation presented here can provide us with an explicit dependence for $D(r)$ and thus lead us to estimates of the flow, given by Eq. (2.66), which was the crucial quantity in the applications discussed in the previous Sections.

If D is used to represent the constant diffusion coefficient used in the Smoluchowski equation, which satisfies the Einstein fluctuation and dissipation theorem, and if Φ_{DE} is the flow calculated by the Smoluchowski equation, one can define

$$D(r) = - \Phi_{DE} \left(\frac{dc}{dr} \right)^{-1}$$

from which, on substitution, one obtains

$$D(r) = D \left\{ 1 + 3 \exp\left[-\ell(r-R)\right] \frac{A(r)(1 + 1.85\,\ell R)}{A(R)(1 + 0.85\,\ell r)} \right. [2.9 + \qquad\qquad (2.109)$$

$$+ 1.45 \, \ell r + 0.3 \, (\ell r)^2 + 2/\ell r]\}^{-1} \tag{2.109}$$

The trend of Eq. (2.109), shown in Fig. 2.7, reproduces the inhibition of diffusion close to the surface which also emerges from the numerical simulations and is also a common characteristic of the functions used in current modelling.

The ratio between $\Phi_{DE}(R)$ and the flow calculated by the Fokker-Planck equation

$$\Phi(R) = - D(R) \left. \frac{dc}{dr} \right|_{r=R}$$

leads to

$$s = \frac{\Phi(R)}{\Phi_{DE}(R)} = \frac{4}{3\pi} \frac{\ell R}{C(R)} (1 + 1.85 \, \ell R) \tag{2.110}$$

For $\ell R = 2, 10, 100$ the values obtained are, respectively, $s = 0.39, 0.78$, and 0.96. For the case of ions in sufficiently dilute solutions (where Coulomb interaction may be ignored) typical values of ℓR, at $25\,°C$ are $\ell R = 78, 41$, and 15 for sodium, potassium and iodide ions respectively, and thus s is much less than 1 for the iodide ion.

An encouraging agreement was found between the values of s calculated using Eq. (2.110) and the corresponding semi-empirical values relating to the phenomenon of coagulation in aerosols.

It is also worth noting that there is thus the possibility of creating models of non-perfect absorbers for which the absorption will depend upon the value of the velocity of the particle with respect to the absorber and that these models thus provide a satisfactory alternative to the semi-empirical procedures currently used for modelling situations of imperfect absorption.

3

Theory of non-Markovian Stochastic Processes

3.1. RELEVANT VARIABLES AND NON-MARKOVIAN PROCESSES

Up to now we have dealt with Markovian processes, which, as we recall, are processes in which a knowledge of the present state of the system is sufficient for the statistical determination of its future development. Using a sufficiently large number of variables every physical system is Markovian, as far as we know. However, in some cases, this could mean a number of variables of the order of Avogadro's number, which would clearly be unmanageable.

It is an experimental fact that in numerous physical systems it is possible to individuate a limited subset of variables, such that their evolution is of a Markovian nature. This is the case, for example, with Brownian motion, where the position and velocity of the Brownian particle are sufficient to obtain a Markovian process. It is to be remembered here that only the position of the particle is insufficient, if one excludes the overdamped case.

Therefore, the possibility of using the theory of Markovian stochastic processes developed above for the study of a given physical system depends upon which variables we use to represent that system, amongst the myriad of variables necessary for an accurate microscopic description. The choice of some of these variables of interest (or relevant variables) could be determined by our own requirements, that is, the type of knowledge desired about the behaviour of the system.

If the set of variables of interest follows Markovian laws of evolution, we can directly apply that which we have learnt in Chapter 1. Otherwise, we must enlarge the set of variables describing the system, to regain a Markovian situation. Hereafter these additional variables will be called "virtual" variables.

The choice of the additional variables may be directed by physical considerations about the system. A simple example is that of Brownian motion: if the variable of interest is the position, then the virtual variable will be the velocity of the Brownian particle. In other cases individuating suitable virtual variables may be much more complicated and, as we shall see, there are systematic techniques which can be helpful.

Now, however, we will deal with another aspect of the problem. After having introduced the virtual variables and written the Fokker-Planck equation for the complete system, we must extract the information relating only to the variables of interest. In other words, we must have at our disposal a technique for "projecting" our equations onto the subspace of the variables of interest.

This operation is possible when the dynamics of the virtual variables is significantly faster than that of the variables of interest. For example, let us consider again overdamped Brownian motion, where the virtual variable (the velocity) relaxes much faster than the variable of interest (the position). Fortunately, the questions we are interested in asking about the system often regard the slower variables, and it is thus possible to eliminate the less interesting variables which are also the faster ones. This operation takes the name of "adiabatic elimination", and we have already seen an example in the case of overdamped Brownian motion.

In that particular case we employed a rather crude technique (direct adiabatic elimination), which essentially consists of putting the derivatives with respect to time of the fast variables equal to zero. In this way algebraic equations are obtained which express the fast variables as functions of the slow ones. Substituting the expressions thus obtained in the equations of motion of the latter leads to a system of equations for only the variables of interest.

Briefly, the direct adiabatic elimination procedure is as follows. Let us suppose that we have a system described by a set of N variables $q = (q_1, \ldots q_N)$. These variables obey a set of differential equations of the type

$$\dot{q}_j = h_j(q) \qquad\qquad j = 1, \ldots, N \qquad\qquad (3.1)$$

By means of a suitable transformation of variables $p = p(q)$, it is possible to write these equations in the following way

$$\dot{p}_j = -\gamma_j p_j + q_j(p) \qquad\qquad j = 1, \ldots, N \qquad\qquad (3.2)$$

where the stationary state ($\dot{p}_j = 0, \forall j$) corresponds to the state in which every p_j is zero. The magnitude of the damping coefficient determines which variables are slow and which are fast.

Let us suppose that we can divide them into two groups, indicated respectively by the subscripts s and f. In the equations for the fast variables we then put $\dot{p}_f = 0$, $f = M + 1, \ldots, N$. In this way we obtain algebraic equations of the type

$$p_f = g_f(p)/\gamma_f \qquad\qquad f = M + 1, \ldots, N \qquad\qquad (3.3)$$

in which we can put all the values of p_f in the r.h.s. equal to the equilibrium value, i.e. zero, because their relaxation is much faster than that of the first M variables. In this way we obtain the f variables as algebraic functions of the s variables. By substituting in the first M equations of the system (Eq. (3.2)), we obtain a system of equations for only the slow variables.

All of this regards deterministic systems, but it can easily be generalised to the case of Langevin-type systems. In fact, the manner in which we have treated Brownian motion in the high viscosity limit is itself an example of this kind. Clearly, however, such a drastic procedure can only lead to a first approximation of the dynamics of the variables of interest. In order to describe more accurately non-Markovian systems, more refined projective techniques are necessary such as those examined below.

3.2 PROJECTION METHODS AND THE ADIABATIC ELIMINATION PROCEDURE

Let us subdivide the variables of our system into two sets, the variables of interest, collectively indicated with a, and the irrelevant ones b. Under the same formalism we can consider two different cases:
1) where the set $\{a, b\}$ gives a complete microscopic characterisation of the system and
2) where the set $\{a, b\}$ corresponds to a reduced description, in which the "a" variables are of interest and the "b" variables are virtual ones.

In both cases the time evolution equation for the probability density $p(a, b, t)$ will be of the form

$$\frac{\partial}{\partial t} p(a, b, t) = Lp(a, b, t) \tag{3.4}$$

where L is a differential operator. In case 1), for example, Lp corresponds to the Poisson bracket of p with the Hamiltonian of the system, if the reasoning is based on classical mechanics, whereas it is proportional to the commutator with the Hamiltonian operator in a quantum context. In case 2), however, the above equation takes on the form of a Fokker-Planck equation. Since we are interested only in the "a" variables, we can define the probability density referred only to the latter as

$$s(a, t) = \int db \, p(a, b, t) \tag{3.5}$$

It is also possible to divide the dynamic operator L into three parts, in one of which only the "a" variables appear, in another only the "b" variables, and in the third, the terms that "mix" the two sets of variables:

$$L(a, b, t) - L_a + L_b + L_i \tag{3.6}$$

We will also assume that an equilibrium distribution exists for only the "b" variables, $p_{eq}(b)$, so that

$$L_b p_{eq}(b) = 0$$
$$\int p_{eq}(b) \, db = 1 \tag{3.7}$$

We can now define a projection operator P for only the "a" variables, by means of the expression

$$Pp(a, b, t) = p_{eq}(b) \int db \, p(a, b, t) = p_{eq}(b) s(a, t) \tag{3.8}$$

To be sure that we are really dealing with a projection operator, we must verify that it is idempotent, that is, the property that repeated applications of the operator generate the same result as a single application. A straightforward calculation shows that this is the case; in fact

$$P^2 p(a, b, t) = Pp_{eq}(b) s(a, t) = p_{eq}(b) \int db \, p_{eq}(b) s(a, t) =$$
$$= p_{eq}(b) s(a, t) = Pp(a, b, t)$$

We can then define the projected densities p_1 and p_2 as:

$$p_1 = Pp(a, b, t)$$
$$p_2 = (1 - P) p(a, b, t) \tag{3.9}$$

The symbol 1 indicates the identity operator. Applying the operators P and $Q = 1 - P$ once each to both sides of the fundamental time evolution equation (Eq. (3.4)) we obtain:

$$\frac{\partial p_1}{\partial t} = PL(p_1 + p_2) \tag{3.10a}$$

$$\frac{\partial p_2}{\partial t} = QL\,(p_1 + p_2)$$ (3.10b)

The solution of the second equation reads:

$$p_2(t) = e^{QLt}\,p_2(0) + e^{QLt}\int_0^t e^{-QLs}QLp_1(s)\,ds =$$

$$= e^{QLt}\,p_2(0) + \int_0^t e^{QLs}\,QLp_1(t-s)\,ds$$

Substituting in the first Eq. (3.10), we obtain

$$\frac{\partial}{\partial t}\,p_1(t) = PLPp(t) + PLe^{QLt}\,Qp\,(0) + \int_0^t PLe^{QLs}\,QLPp\,(t-s)\,ds$$ (3.11)

which is a kind of generalised master equation. It has a non-Markovian character, as can be seen by the presence of the memory kernel:

$$\phi(s) = PL\,\exp\,[QL\,s\,]\,QLP$$

The generalised master (or Fokker-Planck) equation (Eq. (3.11)) is the fundamental equation for the developments outlined below. It should be noted that it is an exact equation needing no approximations for its derivation. Its non-Markovian character is a consequence of the procedure of contraction of the variables.

The equation describes the time evolution of only the projected density p_1, and is therefore the equation we were searching for. The influence of the probability density projected over the irrelevant variables, p_2, upon the evolution of p_1, is limited to the effects of the initial conditions (see the second term in the r.h.s. of Eq. (3.11)).

Obviously, our equation is quite complicated and it is not realistic to think that it can be generally solved. If, however, the irrelevant variables have faster dynamics than the variables of interest, we can attempt a perturbative expansion, searching for approximations increasing in accuracy as a function of some "small" parameter.

The form of the master equation which we have obtained is not very suitable for such a perturbative expansion and therefore it is worthwhile changing the variables. This change of variables is exactly analogous to passing to the interaction representation in quantum mechanics, considering Eq. (3.11) as an equation in the Schrödinger representation. In this context, a Langevin equation is analogous to the Heisenberg representation. Naturally, this relationship is a formal one, as we are dealing with classical quantities.

The change of variables is:

$$\rho = e^{-L_0 t}\,p\,(a,\,b,\,t)$$

$$L_0 = L_a + L_b$$ (3.12)

$$L_i(t) = e^{-L_0 t}\,L_i\,e^{L_0 t}$$

The usefulness of this change of variables derives from the fact that it is possible to immediately demonstrate, directly, that ρ obeys a time evolution equation that has the same form as Eq. (3.4), in which, however, the time evolution operator is given by only the intereaction part, $L_i(t)$:

$$(\partial/\partial t)\,\rho(t) = L_i(t)\,\rho(t)$$ (3.13)

As it was possible to obtain a generalised master equation from the time evolution equation (Eq. (3.4)), so it is possible to obtain a time evolution equation for the projected density from Eq. (3.13). The only difference is due to the fact that $L_i(t)$ does not commute with $L_i(t')$ when $t \neq t'$. It is thus necessary to introduce the concept of time ordered exponentials defined as:

$$\overleftarrow{\exp}\left[-\int_0^t A(s)ds\right] = 1 + \sum_{r=1}^{\infty} \frac{(-1)^r}{r!} \int_0^t ds_1 \int_0^{s_1} ds_2 \ldots \int_0^{s_{r-1}} ds_r A(s_1)\ldots A(s_r) \quad (3.14)$$

The master equation that we then obtain is:

$$\frac{\partial}{\partial t} P\rho(t) = PL_i(t)P\rho(t) + PL_i(t) \overleftarrow{\exp}\left[\int_0^t QL_i(\tau)d\tau\right]Q\rho(0) +$$
$$+ \int_0^t PL_i(t) \overleftarrow{\exp}\left[\int_\tau^t QL_i(s')ds'\right]QL_i(\tau)P\rho(\tau)d\tau \quad (3.15)$$

Now, if the irrelevant variables are also the faster ones, it is worth trying a series expansion of the ordered exponential. In fact if the two sets of variables each follow their own dynamics, with an interaction which is not insignificant but of a moderate intensity, then $L_i(t)$ can be thought of as a "small" quantity.

The perturbative technique, based upon the series expansion of the ordered exponential, which appears in Eq. (3.15), is known as the adiabatic elimination procedure. We will discuss this procedure in the following Sections of this Chapter. For a more detailed treatment of this topic we refer the reader to the works of P. Grigolini quoted in the Bibliography. Our treatment strictly follows that of Grigolini and co-workers without, however, being quite so detailed.

3.3 THE LINEAR CHAIN

Let us now apply the adiabatic elimination procedure to a complex system, with the aim of building a reduced model of it in terms of a single relevant variable. The most interesting aspect is the resulting equation which, for the linear system being considered, takes the form of a Fokker-Planck equation. In this way we arrive at the Fokker-Planck equation as the result of a projection on the set of relevant variables. The model examined here consists of an infinitely long chain of particles, of mass m_i, each of which is linearly coupled to the two adjacent particles by ideal springs having elastic constants K_i, K_{i+1}. M is the mass of the first body of the chain and K the elastic constant of the first spring. The Newton equations for the system are then:

$$
\begin{aligned}
\dot{x} &= v & M\dot{v} &= K(x - y_1) \\
\dot{y}_1 &= w_1 & m_1\dot{w}_1 &= -K(y_1 - x) - K_1(y_1 - y_2) \\
\dot{y}_2 &= w_2 & m_2\dot{w}_2 &= -K_1(y_2 - y_1) + K_2(y_3 - y_2) \\
\end{aligned}
$$
$$\ldots \ldots \ldots \ldots \ldots \ldots \ldots \ldots \ldots \ldots \ldots$$
$$
\begin{aligned}
\dot{y}_i &= w_i & m_i w_i &= -K_i(y_i - y_{i-1}) + K_{i+1}(y_{i+1} - y_i)
\end{aligned}
\quad (3.16)
$$

where x, v, y_i and w_i represent the coordinates and the velocities of the particles of mass M and m_i respectively. Using the inter-particle distances:

$$\Delta_0 = x - y_1 \;; \qquad \Delta_1 = y_1 - y_2 \;; \qquad \ldots \qquad \Delta_i = y_i - y_{i+1}$$

Eqs. (3.16) may be rewritten as follows:

$$\dot{x} = v \qquad\qquad \dot{v} = -\frac{K}{M}\Delta_0$$

$$\dot{\Delta}_0 = v - w_1 \qquad \dot{w}_1 = \frac{K}{m_1}\Delta_0 - \frac{K_1}{m_1}\Delta_1$$

$$\cdots\cdots\cdots\cdots \qquad\qquad \cdots\cdots\cdots\cdots\cdots$$

$$\dot{\Delta}_i = w_i - w_{i+1} \qquad \dot{w}_i = \frac{K_{i-1}}{m_i}\Delta_{i-1} - \frac{K_i}{m_i}\Delta_i$$

$$(3.17)$$

The probability density function for the state of the system $p(v, \Delta_0, \Delta_1, \ldots, w_1, w_2, \ldots, t)$ has a time evolution which is described by a Liouville equation

$$\frac{\partial}{\partial t}p(v, \Delta, w, t) = Lp(v, \Delta, w, t) \tag{3.18}$$

where the differential operator L is defined, in agreement with Eqs. (3.17), as

$$L = \frac{K}{M}\Delta_0\frac{\partial}{\partial v} - (v - w_1)\frac{\partial}{\partial\Delta_0} - \left(\frac{K}{m_1}\Delta_0 - \frac{K_1}{m_1}\Delta_1\right)\frac{\partial}{\partial w_1} +$$

$$- (w_1 - w_2)\frac{\partial}{\partial\Delta_1} - \left(\frac{K_1}{m_2}\Delta_1 - \frac{K_2}{m_2}\Delta_2\right)\frac{\partial}{\partial w_2} - \ldots \tag{3.19}$$

Referring back to Section 1.1 it is worth remembering that the Liouville equation is a continuity equation in phase space for the "gas" made up of the representative points of the Gibbs ensemble. In the continuity equation (Eq. (2.86)) the flow J is

$$J = V p(v, \Delta, w, t)$$

where V is the velocity in phase space (and thus a vector of components $\dot{v}, \dot{\Delta}_0, \dot{w}_1, \dot{\Delta}_1, \ldots$). Given that

$$\mathrm{div}\, Vp = \mathrm{grad}\, p \cdot V + p\, \mathrm{div}\, V$$

and also that the velocity field V has zero divergence (as the points of the Gibbs ensemble form an incompressible fluid), substituting in Eq. (2.86) we obtain

$$\frac{\partial}{\partial t}p(v, \Delta, w, t) = -\mathrm{grad}\, p(v, \Delta, w, t) \cdot V$$

which corresponds to Eqs. (3.18) and (3.19). Let us now "cleave" the operator L (Eq. (3.19)) according to Eq. (3.6) taking v as a relevant variable and the remainder as irrelevant variables. Thus

$$L_a = 0 \tag{3.20a}$$

$$L_i = \frac{K}{M}\Delta_0\frac{\partial}{\partial v} - v\frac{\partial}{\partial\Delta_0} \tag{3.20b}$$

$$L_b = w_1 \frac{\partial}{\partial \Delta_0} - \frac{K}{m_1} \Delta_0 \frac{\partial}{\partial w_1} + \frac{K_1}{m_1} \Delta_1 \frac{\partial}{\partial w_1} - w_1 \frac{\partial}{\partial \Delta_1} + \tag{3.20c}$$

$$+ w_2 \frac{\partial}{\partial \Delta_1} - \frac{K_1}{m_2} \Delta_1 \frac{\partial}{\partial w_2} + \frac{K_2}{m_2} \Delta_2 \frac{\partial}{\partial w_2} - \ldots$$

Given the simplicity of the system it is worth giving, at least in this case, a more precise idea of the essential steps relating to the calculation of the first term of the perturbative expansion deriving from Eq. (3.15). The following description is not, however, necessary for an understanding of the following Sections and thus the reader who is not interested in the more detailed treatment may go directly to the result shown in Eq. (3.32).

In order to build the projection operator P needed for the contraction of the overall information contained in Eq. (3.18) and to carry out the analysis of the relevant variable we take $p_{eq}(\Delta, w)$, which is to be substituted into Eq. (3.8), as

$$p_{eq}(\Delta, w) = N \prod_{i=1}^{\infty} \exp\left(-\Delta_{i-1}^2 / 2 \langle \Delta_{i-1}^2 \rangle_{eq}\right) \exp\left(-w_i^2 / 2 \langle w_i^2 \rangle_{eq}\right) \tag{3.21}$$

(where N is a normalisation constant). It is to be noted that Eq. (3.21) represents the equilibrium distribution for the irrelevant variables. Let us now examine some useful properties of the operator P, which we will use later:

$$P \exp(\pm L_b t) = \exp(\pm L_b t) P = P \tag{3.22a}$$

$$[P, \exp(L_a t)] = 0 \tag{3.22b}$$

$$[\partial/\partial t, P] = 0 \tag{3.22c}$$

Here, as usual, square brackets denote the commutator of the two operators enclosed. Eqs. (3.22b) and (3.22c) are obvious while Eq. (3.22a) ensues from the following additional hypotheses. First of all, the operator L_b must be reducible, eventually upon integration by parts over the irrelevant variables, to a linear combination (with constant coefficients) of partial derivatives with respect to the irrelevant variables. Secondly, the condition

$$L_b p_{eq}(\Delta, w) = 0 \tag{3.22d}$$

must be met. Also, recalling that L_b governs the derivative with respect to time of the distribution of the irrelevant variables whenever isolated (i.e. subtracted from the interaction with the relevant variables), it may be concluded that Eq. (3.22d) is equivalent to

$$\frac{\partial}{\partial t} p_{eq}(\Delta, w) = 0 \tag{3.23}$$

which is satisfied as Eq. (3.21) is an equilibrium distribution. A glance at Eq. (3.20c) confirms that in our case Eqs. (3.22) are valid. It is now necessary to transform Eq. (3.15) into a form which is more useful for the calculations. Using Eqs. (3.9), (3.12), (3.22a) and (3.22c) it can be seen that

$$\frac{\partial p_1}{\partial t} = P L_a p + P e^{L_a t} \frac{\partial}{\partial t} \rho$$

On substituting Eq. (3.8) and using Eqs. (3.22b) and (3.22c) one obtains

$$\frac{\partial}{\partial t} s(v,\ t) = L_a s(v,\ t) + p_{eq}^{-1}(\Delta,\ w)\, e^{L_a t}\, \frac{\partial}{\partial t}\, P\, \rho \qquad (3.24)$$

At this point $\partial P\rho/\partial t$ can be obtained from Eq. (3.24) and substituted into the l.h.s. of Eq. (3.15), yielding an integro-differential equation for $s(v,\ t)$:

$$\frac{\partial}{\partial t} s(v,\ t) = L_a s(v,\ t) + p_{eq}^{-1}(\Delta,\ w)\, e^{L_a t}\, P L_i(t)\, P\, \rho\,(t) +$$

$$+ p_{eq}^{-1}(\Delta,\ w)\, e^{L_a t} \int_0^t P L_i(t)\, \overleftarrow{\exp}\left[\int_\tau^t Q L_i(s')\, ds'\right] Q L_i(\tau) P \rho(\tau)\, d\tau +$$

$$+ p_{eq}^{-1}(\Delta,\ w)\, e^{L_a t}\, P L_i(t)\, \overleftarrow{\exp}\left[\int_0^t Q L_i(s')\, ds'\right] Q\, \rho(0)$$

On simplifying $\exp(L_a t)$ in the second term of the r.h.s. and using Eqs. (3.22a) and (3.22b) again, the foregoing equation becomes:

$$\frac{\partial}{\partial t} s(v,\ t) = L_a s(v,\ t) + p_{eq}^{-1}(\Delta,\ w)\, P L_i P\, e^{L_a t}\, \rho(t) +$$

$$+ \int_0^t K(t-\tau)\, s(v,\ \tau)\, d\tau + \qquad (3.25)$$

$$+ p_{eq}^{-1}(\Delta,\ w)\, e^{L_a t} P L_i(t)\, \overleftarrow{\exp}\left[\int_0^t Q L_i(s')\, ds'\right] Q\, \rho(0)$$

having defined

$$K(t-\tau) = p_{eq}^{-1}(\Delta,\ w)\, P L_i\, e^{L_0 t}\, \overleftarrow{\exp}\left[\int_\tau^t Q L_i(s')\, ds'\right] Q L_i(\tau) e^{-L_0\tau}\, p_{eq}(\Delta,\ w) \qquad (3.26)$$

The above integral can be considerably simplified by assuming that the memory kernel $K(t-\tau)$ falls to zero over periods of time which are very short with respect to the evolution period of the function $s(v,\ \tau)$. In this case, in fact, $s(v,\ \tau)$, which can be considered as a series expansion around the time t, can be approximated by $s(v,\ t)$. The inclusion of successive terms leads, in the perturbative expansion, to contributions of a higher order which will not be taken into consideration here. Moreover, if in the integral

$$\int_0^t K(t-\tau)\, s(v,\ t)\, d\tau$$

the kernel $K(t-\tau)$ has a memory limited to times very close to t, the lower limit can be extended to infinity without appreciable changes in the result so that, on turning to the variable $\alpha = t - \tau$, one has

$$\int_{-\infty}^t K(t-\tau) s(v,\ t)\, d\tau = \int_0^{+\infty} s(v,\ t)\, K(\alpha)\, d\alpha \qquad (3.27)$$

Finally, one can use the assumption that the interaction, and thus L_i, is small enough to allow the utilisation of only the first terms of the expansion (Eq. (3.14)). Thus, in order to calculate only the lowest order terms in the resulting perturbative expansion, it will suffice to substitute into Eq. (3.26)

$$\overleftarrow{\exp}\left[\int_\tau^t QL_i(s')\,ds'\right] \approx 1$$

Regarding the term relating to the initial condition in Eq. (3.24), let us assume a factorised condition corresponding to a situation of equilibrium for all the variables

$$\rho(v, \Delta, w, 0) = p(v, \Delta, w, 0) \propto \exp\left(-v^2/2 \langle v^2 \rangle_{eq}\right) P_{eq}(\Delta, w)$$

In this case it can be directly verified that

$$Q\,\rho(v, \Delta, w, 0) = 0$$

so that the fourth term of Eq. (3.25) also vanishes. Taking into account all of the points made so far and recalling that $L_a = 0$, Eq. (3.24) can be rewritten as

$$\frac{\partial}{\partial t} s(v, t) = p_{eq}^{-1}(\Delta, w) P L_i P \rho(v, \Delta, w, t) +$$

$$+ p_{eq}^{-1}(\Delta, w) \int_0^\infty P L_i e^{L_b t} Q e^{-L_b(t-\alpha)} L_i P_{eq}(\Delta, w) s(v, t)\,d\alpha \qquad (3.28)$$

Regarding the first term of Eq. (3.28), it can be seen that

$$P L_i P = 0 \qquad (3.29)$$

since integrals of the following type are produced

$$\int_{-\infty}^{+\infty} y\, e^{-\beta y^2}\,dy$$

and thus Eq. (3.28) is reduced, at the lowest perturbative order, to

$$\frac{\partial}{\partial t} s(v, t) = \int_0^\infty p_{eq}^{-1}(\Delta, w) P L_i e^{L_b t} Q e^{-L_b(t-\alpha)} L_i s(v, t) P_{eq}(\Delta, w)\,d\alpha \qquad (3.30)$$

Given the definition of Q and the commutation rule (3.22a), we can use Eq. (3.29) again to further simplify Eq. (3.30) to:

$$\frac{\partial}{\partial t} s(v, t) = \int_0^\infty p_{eq}^{-1}(\Delta, w) P L_i e^{L_b \alpha} L_i s(v, t) p_{eq}(\Delta, w)\,d\alpha \qquad (3.31)$$

The structure of Eq. (3.31) immediately becomes clear on examining the composition of L_i and L_b. For the L_i factor on the left, only the first term is to be taken into account: the second, in fact, generates terms of the $\partial/\partial\Delta_0$ type which, in the integral over Δ_0, clearly vanish (as long as $P_{eq}(\Delta, w)$ vanishes fairly rapidly at infinity). Eq. (3.31) then becomes:

$$\frac{\partial}{\partial t} s(v, t) = \int_0^\infty p_{eq}^{-1}(\Delta, w) P \frac{K}{M} \frac{\partial}{\partial v} e^{L_b(t-\tau)} \left[\frac{K}{M} \Delta_0 \frac{\partial}{\partial v} - v \frac{\partial}{\partial\Delta_0}\right] s(v, t) \cdot$$

$$\cdot P_{eq}(\Delta, w) \, d(t - \tau) =$$

$$= \frac{\partial^2}{\partial v^2} \left[\int_0^\infty P_{eq}^{-1}(\Delta, w) \frac{K^2}{M^2} P \Delta_0 \, e^{Lb(t-\tau)} \Delta_0 P_{eq}(\Delta, w) \, d(t - \tau) \right] s(v, t) +$$

$$- \frac{\partial}{\partial v} v \left[P \int_0^\infty P_{eq}^{-1}(\Delta, w) \frac{K}{M} \Delta_0 \, e^{Lb(t-\tau)} \frac{\partial}{\partial \Delta_0} P_{eq}(\Delta, w) \, d(t - \tau) \right] s(v, t) = \tag{3.32}$$

$$= \frac{\partial^2}{\partial v^2} \left[D \, s(v, t) \right] - \frac{\partial}{\partial v} \left[\beta v \, s(v, t) \right]$$

which clearly possesses a Fokker-Planck-type structure. The formal analogy with Eq. (1.52) is further reinforced because the coefficients D and β of Eq. (3.32) can be shown to satisfy a fluctuation-dissipation relationship similar to that of Eq. (1.59). In fact, recalling that

$$P_{eq}^{-1}(\Delta, w) P = P P_{eq}^{-1}(\Delta, w) = \int d\Delta \, dw$$

and

$$\frac{\partial}{\partial \Delta_0} P_{eq}(\Delta, w) = -\frac{\Delta_0}{\langle \Delta_0^2 \rangle_{eq}} P_{eq}(\Delta, w)$$

(see Eqs. (3.8) and (3.21)), it follows that

$$D = \frac{\beta K \langle \Delta_0^2 \rangle_{eq}}{M}$$

This coincides with Eq. (1.59) provided that the equipartition theorem

$$\frac{1}{2} K \langle \Delta_0^2 \rangle_{eq} = \frac{1}{2} k_B T$$

is taken into account (k_B is the Boltzmann constant and T is the effective temperature of the linear chain acting upon M as a thermal bath).

Thus, the overall effect of the masses m_i upon the motion of the mass M is a Brownian motion effect. Naturally, Eq. (3.32) assumes a sufficient separation of the time scales characteristic of the motion of M and of the m_i. This means that, assuming that the elastic constants K and K_i are of the same order of magnitude, $m_i \ll M$, which is the usual order relation between the Brownian particle and the particles of the medium. Calculation of the successive terms of the perturbative expansion shows that their inclusion does not alter the structure of Eq. (3.32) but leads only to a progressive renormalisation of the coefficients.

3.4 ADIABATIC ELIMINATION AND MULTIPLICATIVE NOISE

We will now give a further example of the adiabatic elimination procedure described in the previous Sections. This will serve to clarify the ideas already discussed and also help to shed some light on an important problem in the modelling of random physical systems. We will therefore make a brief digression on the topic of multiplicative noise.

Up to this point, we have met Langevin equations of the following type:

$$\dot{x} = f(x) + A(t) \tag{3.33}$$

where $A(t)$ is a delta-correlated white force. In some cases, however, as we have seen in Chapter 1, it is possible to employ stochastic models of the type

$$\dot{x} = f(x) + g(x) A(t) \tag{3.34}$$

where the (white) noise term is multiplied by a function of the random variable x.

This type of equation presents new problems; in particular, as we shall see, its meaning is ambiguous and different precepts exist in the literature, associated with the names of Itô and Stratonovich, for associating a Fokker-Planck equation with Eq. (3.34).

The ambiguity is linked to the "pathological character" of the functions of a stochastic process, which we can understand by considering, for simplicity, a function $g(w)$ where $w(t)$ is a Wiener process. Applying the usual rules of calculus for the differential dg we would write the expression:

$$dg(w) = g'(w) \, dw \tag{3.35}$$

Generally, when we operate with stochastic variables, at a certain point we are interested in carrying out averaging operations. Averaging the previous equation, however, we obtain:

$$\langle dg(w) \rangle = \langle g'(w) \, dw \rangle = \langle g'(w) \rangle \langle dw \rangle \tag{3.36}$$

where the last step is a consequence of the fact that Brownian motion is a process with independent increments. Since $\langle dw \rangle = 0$, we obtain $\langle dg \rangle = 0$, and we have thus lost the possibility of following the time evolution of our variable.

The error committed is that of applying the usual rules of infinitesimal calculus, forgetting that in the case of Brownian motion $\langle dw^2 \rangle$ is of the same order as dt. It is therefore incorrect to stop at the first order of dw, and the rule expressed by Eq. (3.35) must be replaced by:

$$dg(w) = g'(w) \, dw + (1/2) g''(w) \, dw^2 \tag{3.37a}$$

As we are interested in average values, it is possible to substitute dw^2 with $2D dt$ (where D is the diffusion coefficient):

$$dg(w) = g'(w) \, dw + D g''(w) dt \tag{3.37b}$$

The changes made in the rules of differentiation also apply, in an analogous way, to the rules of integration; if G is a primitive of g, from Eqs. (3.37a) and (3.37b) we obtain

$$
\begin{aligned}
\int_{w_A}^{w_B} g(w) \, dw &= G(w_B) - G(w_A) - (1/2) \int_{w_A}^{w_B} g'(w) \, dw^2 = \\
&= G(w_B) - G(w_A) - D \int_{t_A}^{t_B} g' \, dt
\end{aligned}
\tag{3.38}
$$

These rules of calculus are known as "Itô's rules". Within the theory of stochastic differential equations, the conventional rules of calculus are known as the "Stratonovich rules". Using the symbol $\circ$ for the calculus according to Stratonovich, we have

$$\int_{w_A}^{w_B} g(w) \circ dw = G(w_B) - G(w_A) \tag{3.39}$$

The relationship between the integrals of Itô and Stratonovich is therefore:

$$\int_{w_A}^{w_B} g(w)\, dw = \int_{w_A}^{w_B} g(w) \circ dw - D \int_{t_A}^{t_B} g'\, dt \tag{3.40}$$

It is to be noted that in the case of additive noise (Eq. (3.33)), the two expressions coincide, whereas they are different in the case of multiplicative noise described by Eq. (3.34).

We recall that the differential notation employed in Langevin-type equations constitutes a kind of abbreviated notation, lack mathematical rigour due to the presence of the white force. Eq. (3.34) is therefore nothing other than a stenographic form of the more correct form:

$$x(t) - x(0) = \int_0^t f(x(s))\, ds + \int_0^t g(x(s))\, dw(s) \tag{3.41}$$

An interpretation of the Langevin equation, however, requires a choice on how to interpret the stochastic integral which appears in the r.h.s. of Eq. (3.41).

The considerations so far would favour the choice of Itô, but the question is, in reality, more complicated. In fact, since only Eq. (3.41) has a rigorous meaning, both interpretations are possible and lead to results which are consistent from a mathematical point of view.

In reality, we are dealing with a physical problem: some systems might be adequately represented by a Langevin equation of the kind given by Eq. (3.34) interpreted according to Itô, and others by the same Langevin equation interpreted according to Stratonovich.

If we adopt Itô's interpretation, the transition density $p(x, t \mid x_0)$ obeys the following Fokker-Planck equation:

$$\frac{\partial}{\partial t} p(x, t \mid x_0) = - \frac{\partial}{\partial x} (fp) + D \frac{\partial^2}{\partial x^2} (g^2 p) \tag{3.42}$$

On the other hand, following Stratonovich, the corresponding Fokker-Planck equation is:

$$\frac{\partial}{\partial t} p(x, t \mid x_0) = - \frac{\partial}{\partial x} (fp) + D \frac{\partial}{\partial x} \left[g \frac{\partial}{\partial x} (gp) \right] \tag{3.43}$$

However, on the basis of physical intuition, we are often led to write a Langevin-type equation instead of a master or a Fokker-Planck equation, for whose interpretation there would be no uncertainty. Intuition, on the other hand, can not be used for choosing between the two interpretations, so that the problem remains unsolved.

Without attempting to be rigorous we shall mention two theoretical results which could be of help in choosing the more suitable interpretation. The first refers to the case in which the multiplicative white noise $g(x) A(t)$ can be considered as the limit of a succession of terms of the type $g(x) A_K(t)$, where the $A_K(t)$ are Gaussian functions which tend towards white noise, or rather

$$\langle A_K(t) A_K(s) \rangle = 2D\, \phi_K(t - s)$$

$$\lim_{K \to \infty} \phi_K(t - s) = \delta(t - s)$$

In this case, the above theorem affirms that the correct interpretation is that of Stratonovich. The second theorem affirms that, if our Langevin equation originates from the continuous limit of a discrete Markov chain, then we must adopt Itô's rules.

Let us pause for a moment to discuss the origin of the multiplicative noise term. One possible cause is the interaction with the external environment: this can, in fact, cause one of the parameters of our equations to fluctuate. One example is that of a chemical reaction in which one of the intermediates is light-sensitive. Illuminating the reaction vessel with a fluctuating light source that tends towards white noise (in other words, a source having a flat spectrum over a wide frequency interval), we can realise the case of a parameter which fluctuates due to the influence of the environment.

There is also another possible origin of multiplicative noise which leads us back to the adiabatic elimination of variables. Let us suppose that we have defined the boundaries of our system so that all the significant sources of noise are included, and that we can thus ignore the multiplicative noise generated by interactions with the external environment. It can still happen that additive white noise terms become multiplicative following the adiabatic elimination of some of the variables.

We can illustrate this with an example of a reduced model. Let us consider:

$$\dot{x} = v$$

$$\dot{v} = -\lambda v - \frac{dU(x)}{dx} + g(x)\,\xi \tag{3.44}$$

$$\dot{\xi} = -\gamma \xi + \eta(t)$$

where $\eta(t)$ is a white force with a diffusion coefficient D. This model is quite interesting from the applicative point of view. Whilst deferring a discussion of its physical meaning until the next Section, it is worth mentioning that, for the case in which $U(x)$ is a double-well potential, this system can describe chemical reactions in the presence of an oscillating potential barrier. This fact could find a certain relevance to the discussion of catalytic phenomena.

We will now illustrate the origin of multiplicative noise, adiabatically eliminating, in a direct way, the virtual variables v and ξ. We then obtain:

$$\dot{x} = \frac{1}{\lambda}\frac{dU}{dx} + \frac{1}{\lambda\gamma}g(x)\,\eta(t) \tag{3.45}$$

which has the form of Eq. (3.34). We still have the problem of choosing between the different interpetations of Itô and Stratonovich. While direct adiabatic elimination can not help us in this choice, the adiabatic elimination procedure, as we shall see, is quite illuminating. From the model equations (Eq. (3.44)) we can directly deduce the form of the Fokker-Planck equation for the transition density $p(x, v, \xi, t \mid x_0, v_0, \xi_0)$. This is

$$\frac{\partial p}{\partial t} + \frac{\partial}{\partial x}(v\,p) + \frac{\partial}{\partial v}\left[\left(-\lambda v - \frac{dU}{dx} + g(x)\,\xi\right)p\right] + \frac{\partial}{\partial \xi}(-\gamma\xi p) = D\frac{\partial^2 p}{\partial \xi^2} \tag{3.46}$$

The Fokker-Planck equation can be written in the form:

$$\frac{\partial}{\partial t}p = Lp$$

$$L = L_0 + L_i$$

$$L_0 = \lambda\frac{\partial}{\partial v}v + \gamma\frac{\partial}{\partial \xi}\xi + D\frac{\partial^2}{\partial \xi^2} \tag{3.47}$$

$$L_i = -v\frac{\partial}{\partial x} + \left[\frac{dU}{dx} + g(x)\,\xi\right]\frac{\partial}{\partial v}$$

where we have employed the same notation as in Section 3.3, considering x as the variable of interest. We now project on x by applying the operator P defined as:

$$Pp = \rho_{eq}(v)\,\rho_{eq}(\xi)\,s(x, t)$$

$$s(x, t) = \int dv\,d\xi\, p(x, v, \xi, t)$$

(3.48)

where $\rho_{eq}(v)$ and $\rho_{eq}(\xi)$ are the equilibrium distributions of the two virtual variables, and where, for the sake of simplicity, we have ignored the dependence of p upon the initial point, considering it as having a fixed value.

This leads us back to the formalism of Section 3.3, and we clearly must proceed by changing to the interaction picture and developing the ordered exponentials. We will not reproduce here the laborious calculation necessary, giving only the expression obtained by including up to fourth-order terms in the operator $L_i(t)$ (a more detailed treatment of this example can be found in the original papers of Grigolini's group, cited in the bibliographical notes to this Chapter):

$$\frac{\partial}{\partial t} s(x, t) = \left\{ \frac{1}{\lambda}\frac{\partial}{\partial x}\frac{dU}{dx} - \frac{1}{\lambda^3}\frac{\partial}{\partial x}\frac{dU}{dx}\frac{d^2 U}{dx^2} + \right.$$
$$\left. + \frac{D}{\lambda^2\gamma^2}\frac{\partial^2}{\partial x^2} g^2(x) - \frac{D}{\lambda\gamma^2(\lambda+\gamma)}\frac{\partial}{\partial x} g(x)g'(x) \right\} s(x, t)$$

(3.49)

In order to clarify the limits of validity of the possible choices regarding the type of stochastic calculus, we will compare Eq. (3.49) obtained by a perturbative development of the exact Fokker-Planck equation (Eq. (3.46)) with the Fokker-Planck equations obtained according to Itô and Stratonovich starting from the Langevin equation for only x (Eq. (3.45)).

These are respectively (see the rules furnished by Eqs. (3.42) and (3.43)):

$$\frac{\partial}{\partial t} s(x, t) = \left\{ \frac{1}{\lambda}\frac{\partial}{\partial x}\frac{dU}{dx} + \frac{D}{\lambda^2\gamma^2}\frac{\partial^2}{\partial x^2} g^2(x) \right\} s(x, t)$$

(3.50)

according to Itô and

$$\frac{\partial}{\partial t} s(x, t) = \left\{ \frac{1}{\lambda}\frac{\partial}{\partial x}\frac{dU}{dx} + \frac{D}{\lambda^2\gamma^2}\frac{\partial}{\partial x} g(x)\frac{\partial}{\partial x} g(x) \right\} s(x, t)$$

(3.51)

according to Stratonovich.

It can be seen that neither of the two recipes reproduces the time evolution described by Eq. (3.49). We can also see that the adiabatic elimination only makes sense if v and ξ are overdamped variables, i.e. if λ and γ are "large" parameters.

Let us first consider the case in which λ and γ are both large, and also where $\lambda \gg \gamma$. We can thus ignore the term in λ^{-3} in Eq. (3.49); besides this, if $\lambda + \gamma \sim \lambda$, the sum of the diffusive terms becomes

$$\frac{D}{\lambda^2\gamma^2}\frac{\partial^2}{\partial x^2}(g^2 s) - \frac{D}{\lambda\gamma^2(\lambda+\gamma)}\frac{\partial}{\partial x}(gg's) \simeq \frac{D}{\lambda^2\gamma^2}\frac{\partial}{\partial x}\left\{ g\frac{\partial}{\partial x}(gs) \right\}$$

Thus, in this limit, we again obtain exactly the same result as Stratonovich.

Let us now consider the case in which $\lambda \ll \gamma$. Then $\lambda + \gamma \sim \gamma$ and the sum of the diffusive terms of Eq. (3.49) becomes:

$$\frac{D}{\lambda^2 \gamma^2} \left\{ \frac{\partial^2}{\partial x^2} (g^2 s) - \frac{\lambda}{\gamma} \frac{\partial}{\partial x} (g g' s) \right\} \simeq \frac{D}{\lambda^2 \gamma^2} \frac{\partial^2}{\partial x^2} (g^2 s)$$

Therefore, in this case Itô's result gives a good approximation for the diffusive terms of the time evolution equation for the probability density.

It is worth noting that in this case it is no longer possible to simply ignore the λ^{-3} with respect to the $(\lambda \gamma)^{-2}$ term. In any case this takes the form of a deterministic term, and can be interpreted as a correction to the potential $U(x)$; since λ is always large, the correction will be small with respect to the dominating term. Therefore, in this case, Itô's rules give an approximately valid description of our system.

These considerations can be summarised by stating that it is the relative rate of relaxation of the fast variables which determines which type of stochastic calculus is the more appropriate.

If it is possible to give a physical significance to the virtual variables, it will be possible to formulate hypotheses about their relative dynamics, and therefore to choose the correct type of stochastic calculus on a physical instead of a formal basis.

3.5 FLUCTUATING POTENTIAL BARRIERS

We would now like to apply the adiabatic elimination procedure to a model which can be considered as a generalisation of Kramers' model, and which can therefore find applications in chemical kinetics and other sectors. The model is defined by the following equations:

$$\begin{cases} \dot{x} = v \\ \dot{v} = -\lambda v - \dfrac{d}{dx} U(x) + g(x)\,\xi + f(t) \\ \dot{\xi} = -\gamma \xi + \eta(t) \end{cases} \tag{3.52}$$

where $f(t)$ and $\eta(t)$ are white forces having diffusion coefficients D and D_η respectively.

Let us suppose that the potential $U(x)$ has a double-well shape, and is defined as

$$U(x) = U_0 a^{-4} (x^2 - a^2)^2 \tag{3.53}$$

with $U_0 > 0$ (see Fig. 3.1). The deterministic "force" is therefore

$$-dU/dx = -(4 U_0/a^2) x + (4 U_0/a^4) x^3 \tag{3.54}$$

We also assume that the function g is linear:

$$g(x) = \mu x \tag{3.55}$$

The presence of the term $g(x)$ gives rise to an "effective force" $F_{eff}(x)$ such that

$$\dot{v} = -\lambda v + F_{eff}(x, t) + f(t) \qquad F_{eff}(x, t) = \left(\mu \xi - \frac{4 U_0}{a^2} \right) x + (4 U_0/a^4) x^3 \tag{3.56}$$

The effective force, as well as the relative "effective potential", is a random variable as it depends upon the Ornstein-Uhlenbeck process ξ (see the last Eq. (3.52)).

The study of such fluctuations in the potential barrier is important as it may be hypothesized that in certain catalytic reactions, the catalyst-environment interaction could give rise to effects which can be represented in this way. Also, in other physical examples one may have to deal with fluctuating potential barriers (e.g. electrical conduction between metallic islands in a disordered material).

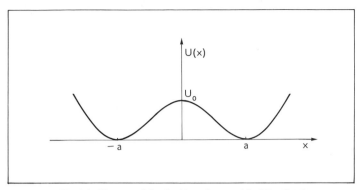

Fig. 3.1 - Diagram of the double-well potential of Eq. (3.53).

In order to apply the results of Section 3.4, we will consider the case in which $f(t)$ is zero. The model, lacking the force f, will serve as a guide for the physical interpretation of the results for the complete model.

Essentially, we can say that the absence of the additive fluctuating force prevents the overcoming of the barrier, since the effective force at the summit is zero. Thus, the transition from one well to the other will in this case be replaced by a migration from one of the wells to the top of the barrier.

Given all this, we can see that the model reduces to that described by Eq. (3.44). Applying the adiabatic elimination procedure, we thus obtain Eq. (3.49) which, by ignoring the term in λ^{-3}, may be written as:

$$\frac{\partial}{\partial t} s(x,\, t) = - \frac{\partial}{\partial x} \{[(d+D)x - bx^3] s(x,\, t)\} + D \frac{\partial^2}{\partial x^2} [x^2 s(x,\, t)] \qquad (3.57)$$

with

$$D = \mu^2 D_\eta /\gamma^2 \lambda^2 \qquad\qquad b = 4U_0/\lambda a^4$$

$$d = (4U_0/\lambda a^2) - D/(1 + R) \qquad\qquad R = \lambda/\gamma \qquad (3.58)$$

The stationary solution of this equation, valid in the half-plane $x > 0$ is :

$$S_{eq}(x) = Qx^\alpha \exp(- bx^2/2D)$$

$$\alpha = (d/2D) - 1 \qquad\qquad (3.59)$$

with $Q = $ constant.

It is to be noted that the probability density (Eq. (3.59)) is positive semidefinite only in the half-plane $x > 0$. This does not cause serious problems since, as we have seen, without the force $f(t)$ there can be no transitions across the barrier. Therefore, considering initial conditions such that all the particles are in this half-plane, they will remain there. The stationary distribution (Eq. (3.59)) is given in Fig. 3.2 (curve (a)).

The abscissa x_M of the maximum and the corresponding value for S_{eq} are defined, for $\alpha > > 0$, as follows:

$$x_M = \sqrt{\frac{\alpha D}{b}} \qquad\qquad (3.60a)$$

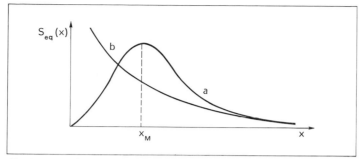

Fig. 3.2 - Plot of the stationary solution of Eq. (3.57); curve (a) and (b) refer respectively to positive and negative values for the parameter α.

$$S_{eq}(x_M) = Q \left(\frac{\alpha D}{b} \right)^{\alpha/2} e^{-\alpha/2} \qquad (3.60b)$$

We can see that on increasing α the maximum moves to the right whereas it tends towards 0 as $\alpha \to 0$. If, however, α is negative, the distribution (Eq. (3.59)) presents a singularity at the origin (see curve (b) in Fig. 3.2).

Therefore, on decreasing α from an initially positive value, the distribution changes shape, its maximum moving towards the point $x = 0$. When it reaches the threshold value $\alpha = 0$, a drastic change in the shape of the distribution occurs, with an increasing particle density at the summit of the barrier which would correspond, in a model containing $f(t)$, to a sharp acceleration in the transition rate.

This sharp change in the shape of the distribution corresponding to a threshold value of the parameter α is an effect of the multiplicative noise term. Here we meet for the first time the phenomenon of the so-called "noise-induced transitions", typical of systems with multiplicative noise, for which there is no deterministic equivalent. Here the multiplicative noise term has been introduced by the elimination of fast variables, whereas in other cases it may arise from the coupling of the system with a rapidly fluctuating environment. We shall turn again to noise-induced transitions in Section 6.8.

From the definitions of Eqs. (3.58) and (3.59), for α we obtain:

$$\alpha = (d/2D) - 1 = \frac{2U_0}{D\lambda a^2} - \frac{1}{2(1+R)} - 1$$

Thus α is a decreasing function of D, which, in its turn, is proportional to D_η. Therefore, we have seen that there can be a drastic increase in the reaction rate (as the distribution changes from shape (a) to shape (b) of Fig. 3.2) with increasing fluctuations of the barrier.

Finally, we note that these conclusions, obtained here with the simplified model (Eq. (3.44)), have been supported by numerical studies carried out on the more complete model (Eq. (3.52)) and also by analog simulations with suitable electronic circuits, carried out by Grigolini's group.

3.6 THE REDUCED MODEL THEORY

So far, we have seen that it is possible to treat non-Markovian systems by introducing new ("virtual") variables. Also, it often happens that the variables of interest exhibit dynamics which are slow with respect to those of the elementary interactions. It is then possible to introduce virtual variables which are faster than the variables of interest (even though they are still slow compared to the time scales at the microscopic level).

In this case the virtual variables (introduced in order to recover a Markovian system) can be eliminated by applying the above adiabatic procedure, thus obtaining a description in terms of only the variables of interest.

The effective utility of this procedure, which goes under the name of "reduced model theory", is based upon the identification of a convenient set of virtual variables. In some cases the physics of the problem suggests which variables should be introduced. However, when intuition does not directly indicate which variables to use, we can obtain some indications from a theoretical approach due to Mori.

Before examining this theory in a general way, we will take a more concrete case in order to better illustrate the basic features of the theory .

The physical example, once again, is Brownian motion, of which we will now give a more complete description than we have presented so far. Up to now, in fact, we have represented the friction with a term $-\gamma v(t)$. In reality, the fluid has a certain inertia, so that the state of the molecules surrounding the Brownian particle at an instant t depends upon the motion of the Brownian particle during the preceding instants. In other words a detailed microscopic knowledge of the whole system at an instant t can be substituted, at least approximately, by a knowledge of the motion of the single Brownian particle during the preceding instants.

The reduced non-Markovian equation for the motion of the particle will then be of the type:

$$\dot{v}(t) = -\int_0^t \varphi(t-s)\, v(s)\, ds \tag{3.61}$$

Eq. (3.61) is a deterministic equation. Its stochastic equivalent is a generalised non-Markovian Langevin equation:

$$\dot{x}(t) = v(t)$$
$$\dot{v}(t) = -\int_0^t \varphi(t-s)\, v(s)\, ds + F(t) \tag{3.62}$$

In this equation $F(t)$ is a stochastic force, the characteristics of which will now be determined. In particular, as we shall demonstrate, it is not a white force, except in the limiting case in which the memory function reduces to a Dirac delta.

We will suppose that the stochastic force has a zero average value and is not correlated with the initial velocity v_0:

$$\langle F(t)\rangle = 0$$
$$\langle v_0 F(t)\rangle = 0 \tag{3.63}$$

Taking the Laplace transform of Eqs. (3.62), we convert them into the following algebraic system (the symbol ^ refers to the Laplace transform of the corresponding time function):

$$z\hat{x}(z) - x_0 = \hat{v}(z) \tag{3.64a}$$

$$z\hat{v}(z) - v_0 = -\hat{\varphi}(z)\,\hat{v}(z) + \hat{F}(z) \tag{3.64b}$$

Solving Eq. (3.64b) for $\hat{v}(z)$, substituting in Eq. (3.64a) and then antitransforming, we obtain:

$$x(t) = x_0 + v_0\,\psi(t) + \int_0^t \psi(t-s)\,F(s)\,ds$$

$$\tag{3.65}$$

$$v(t) = v_0\,\chi(t) + \int_0^t \chi(t-s)\,F(s)\,ds$$

in which the functions ψ and χ are defined by the following equations (the simbol $\mathscr{L}$ indicates the Laplace transform operator, and $\mathscr{L}^{-1}$ its inverse):

$$\chi(t) = \mathscr{L}^{-1}\left\{\frac{1}{z+\hat{\varphi}(z)}\right\}$$

$$\tag{3.66}$$

$$\psi(t) = \mathscr{L}^{-1}\left\{\frac{1}{z(z+\hat{\varphi}(z))}\right\}$$

We now ask ourselves what is the physical significance of χ and ψ. It is straightforward to verify, by multiplying Eqs. (3.65) by v_0 and averaging, that:

$$\frac{\langle v_0 v(t)\rangle}{\langle v_0^2\rangle} = \chi(t) \tag{3.67}$$

$$\frac{\langle v_0(x(t)-x_0)\rangle}{\langle v_0^2\rangle} = \psi(t) \tag{3.68}$$

Therefore, $\chi(t)$ is essentially the correlation function between the initial velocity and the velocity at the instant t, while $\psi(t)$ has an analogous meaning for the position at the instant t. The physical meaning of the equations obtained is apparent: the memory functions for the velocity and the position are essentially the correlation function of the respective variables with the initial velocity.

Another important relationship links the memory function $S(t)$, which appears in the generalised Langevin equation (Eq. (3.62)), with the correlation function of the stochastic force:

$$Q(t-s) = \langle F(s)\,F(t)\rangle \tag{3.69}$$

Given the importance of this relationship, we will provide a detailed derivation. We see that:

$$\langle (v - v_0\chi)^2\rangle = \int_0^t ds \int_0^t ds'\,\chi(t-s)\chi(t-s')\,Q(s-s') =$$

$$\tag{3.70}$$

$$= \int_0^t d\tau \int_0^t d\tau'\,\chi(\tau)\chi(\tau')\,Q(\tau'-\tau)$$

On the other hand

$$\langle (v - v_0\chi)^2\rangle = \langle v^2\rangle + \langle v_0^2\rangle\chi^2 - 2\chi\langle v_0 v\rangle \tag{3.71}$$

Supposing that the system finds itself at thermal equilibrium, the average values of v^2 and of v_0^2

are identical and equal to kT/m. From Eq. (3.67), we also obtain an expression for $\langle vv_0 \rangle$ which, substituted in Eq. (3.71), gives:

$$\frac{kT}{m}(1-\chi^2(t)) = \int_0^t d\tau \int_0^t d\tau' \chi(\tau)\chi(\tau')Q(\tau-\tau') \tag{3.72}$$

In order to obtain this latter expression for $\langle (v-v_0\chi) \rangle$ we have also taken into account Eq. (3.70). Differentiating with respect to time, and including the generalised Langevin equation to express $v(t)$, we obtain:

$$\frac{d\xi}{dt} = -\frac{2kT}{m}\chi\frac{d\chi}{dt} = -\frac{2kT}{m}\chi\frac{1}{\langle v_0^2 \rangle}\frac{d}{dt}\langle v_0 v(t)\rangle =$$

$$= -\frac{2kT}{m}\frac{\chi}{\langle v_0^2 \rangle}\langle v_0 \dot{v} \rangle = \frac{2kT}{m}\chi(t)\int_0^t \varphi(t-s)\chi(s)\,ds \tag{3.73}$$

where $\xi = \langle (v-v_0\chi)^2 \rangle$.

In order to differentiate the r.h.s. of Eq. (3.70) we must employ Leibniz's formula, since the variable t appears both in the upper integration limit and in the integrand. Defining

$$M(\tau, t) = \chi(\tau) R(\tau, t)$$

$$R(\tau, t) = \int_0^t d\tau' \chi(\tau') Q(\tau'-\tau) \tag{3.74}$$

and applying the Leibniz formula, we have:

$$\frac{d\xi}{dt} = \frac{d}{dt}\int_0^t M(\tau, t)\,d\tau = \int_0^t \frac{\partial}{\partial t}M(\tau, t)\,d\tau + M(t, t) =$$

$$= \int_0^t d\tau\,\chi(\tau)\frac{\partial}{\partial t}R(\tau, t) + \chi(t)R(t, t) = \tag{3.75}$$

$$= 2\chi(t)\int_0^t \chi(\tau)\,Q(t-\tau)\,d\tau$$

Finally, comparing the r.h.s.'s of Eqs. (3.74) and (3.72), we obtain the desired relationship between the memory kernel $\varphi(t)$ and the noise autocorrelation function $Q(t)$:

$$(kT/m)\,\varphi(t) = Q(t) = \langle F(0)F(t) \rangle \tag{3.76}$$

In conclusion, by examining the example of Brownian motion we have succeeded in obtaining a generalised Langevin equation (Eq. (3.62)) and a generalised fluctuation-dissipation theorem (Eq. (3.76)), which links the noise autocorrelation function with the deterministic term of the generalised Langevin equation.

The Mori theory generalises these results.

Let us consider a quantity $A(q, p)$, where the symbols q and p respectively denote the set of the generalised coordinates and conjugate moments which provide a complete microscopic description of our physical system. The time evolution equation for A takes the form:

$$\dot{A}(t) = LA(t) \tag{3.77}$$

where L is a differential operator. In classical mechanics, LA is the Poisson bracket of A with the Hamiltonian; in the quantum case, LA would be proportional to the commutator of A with the Hamiltonian.

The functions $\dot{A}$ and A belong to a functional space in which a scalar product is defined (metric space). For the moment we do not need to go into the definition of this scalar product, as long as it holds the properties which define a scalar product. Mori has demonstrated that it is possible, using formal manipulations, to transform Eq. (3.77) into the form:

$$\dot{A}(t) = -\int_0^t \varphi(t-s) A(s)\, ds + F(t) \tag{3.78}$$

where F is a function with the property (the brackets $\langle \rangle$ indicate the scalar product):

$$\langle A_0 F(t) \rangle = 0 \tag{3.79}$$

This is also linked to the kernel $\varphi(t)$ by the expression:

$$\varphi(t) = \frac{\langle F(0) F(t) \rangle}{\langle A_0^2 \rangle} \tag{3.80}$$

The analogy between these equations and those obtained previously for the case of Brownian motion is evident. We note, however, that the Mori equations are purely formal and exact, since it is not necessary to introduce any approximation in order to derive them from the microscopic equation (Eq. (3.77)).

For a physical interpretation it is necessary to define a suitable scalar product. It is possible to define a scalar product as an equilibrium average, by using the expression:

$$\langle A(p, q) B(p, q) \rangle = \int A(p, q) B(p, q) \rho_{eq}(p, q) dp\, dq \tag{3.81}$$

where $\rho_{eq}(p, q)$ is the equilibrium distribution. In this way the scalar product $\langle AB \rangle$ corresponds to the correlation function. In particular, the condition of orthogonality corresponds to the absence of statistical correlations.

This observation suggests that $F(t)$, which is orthogonal to A_0, be interpreted as a noise term, generated by the degrees of freedom which are omitted from the Mori description, which is the result of a projection operation on the single variable of interest.

The explicit expression for $F(t)$ is

$$F(t) = \exp\left[(1 - P) Lt\right] (1 - P) LA_0 \tag{3.82}$$

where P is the operator which projects on the state A_0, the form of which is given here for completeness (using Dirac's notation)

$$P = \frac{|A_0\rangle \langle A_0|}{\langle A_0 | A_0 \rangle} \tag{3.83}$$

We can see from these equations that the time evolution of F is driven not by the operator L, but by its part $(1 - P)L$ projected on the subspace orthogonal to A_0. This confirms the identification of $F(t)$ as a stochastic term, originating from the contraction operated by projecting on A_0.

We can thus conclude that the Mori equations provide adequate generalisations of the Langevin equation and of the fluctuation-dissipation theorem. These are valid under very general conditions, but they certainly do not constitute an instrument of calculus for deriving a macroscopic evolution. In fact, the formal expressions obtained are too complicated for calculation.

However, this theory can help us in the construction of reduced models, whenever we are trying to formulate models from which, by projecting on the variable of interest, we can obtain an equation having the form of Eq. (3.78).

Mori also demonstrated that the Laplace transform $\hat{\varphi}(z)$ of the memory function $\varphi(t)$ can be cast in the form:

$$\hat{\varphi}(z) = \cfrac{\Delta_1^2}{z + \cfrac{\Delta_2^2}{z + \cfrac{\Delta_3^2}{z + \cfrac{\Delta_4^2}{z + \gamma_4(z)}}}} \tag{3.84}$$

where $\gamma_4(z)$ ensures the prosecution of the development with the same form of the preceding terms.

Summarising, therefore, Mori's theory gives us: the form of the generalised Langevin equation; the corresponding generalisation of the fluctuation-dissipation theorem and the general form of the Laplace transform (if it exists) of the memory kernel.

Now, this information can be used in the construction of a reduced model of our system, that is, in the choice of a limited number of virtual variables. It is now required that the equations of the model, projected on the variable of interest, give rise to an equation which takes the form of Eq. (3.78), with a memory kernel such that its Laplace transform is of the type described by Eq. (3.84), truncated at a certain order.

We will now illustrate these considerations with an example taken from, needless to say, Brownian motion. Let us consider the case of a Brownian particle elastically linked to an equilibrium position. The variable of interest is the position x, and the only virtual variable is the velocity v. It goes without saying that we know how to treat this example but we wish to illustrate the points made above using a familiar example.

The equations of the model are

$$\dot{x} = v$$
$$\dot{v} = -\gamma v - cx + A(t) \tag{3.85}$$

where $A(t)$ is a delta-correlated stochastic force and c is a constant. Taking the Laplace transform of the second equation we get

$$\hat{v}(z) = [v_0 + \hat{A}(z) - c\hat{x}(z)]/(z + \gamma)$$

Antitransforming and then substituting into the first Eq. (3.85), we obtain:

$$dx(t)/dt = v_0 e^{-\gamma t} + \int_0^t e^{-\gamma(t-s)} A(s)\, ds - c \int_0^t e^{-\gamma(t-s)} x(s)\, ds \tag{3.86}$$

which has precisely the form required by Mori's theory.

4

Stochastic Processes and Irreversible Thermodynamics

4.1 OUTLINE OF THE THERMODYNAMICS OF IRREVERSIBLE PROCESSES

Classical thermodynamics could be more properly defined as thermostatics, because it concerns the properties of systems in equilibrium, in which the values of the variables remain constant with respect to time. Even when transformations are considered, in reality equilibrium states which are subject to certain macroscopic constraints are compared with other less constrained states of equilibrium.

The thermodynamics of irreversible processes is concerned with defining general laws regarding real transformations in macroscopic systems, which take place at a finite rate obliging the system to pass through non-equilibrium states.

The fundamental principles of irreversible thermodynamics are given in numerous text-books (see the Bibliography) so that we will only briefly summarise them here, underlining the aspects which will serve us in the discussion of their relationship with the theory of stochastic processes.

Our point of departure is the second law of thermodynamics

$$dS/dt \geqslant 0 \tag{4.1}$$

where S is the total entropy of the system. This relationship is valid for a closed system which exchanges neither energy nor material with the exterior. In the case of non-isolated systems, it is assumed that the total change in entropy is the sum of two terms:

$$dS = d_e S + d_i S \tag{4.2}$$

where $d_e S$ describes the change in entropy caused by interactions with the environment and is of undefined sign; $d_i S$ describes the internal dissipative processes and it is assumed, in analogy with Eq (4.1), that it does not take negative values:

$$d_i S \geqslant 0 \tag{4.3}$$

We can see that, by taking into consideration systems which are not in thermodynamic equilibrium, it is no longer justified to assume spatial homogeneity. Entropy, like other thermodynamic variables, thus becomes a field function, and the localised version of the fundamental principle of Eq. (4.3) affirms that the rate of change of S in the small volume dV is equal to the flow of S through the surface which contains the small volume plus the generative contribution of the irreversible processes within dV. We can therefore write a continuity equation for the entropy:

$$\frac{\partial}{\partial t} s(\vec{x}, t) = -\vec{\nabla} \cdot \vec{J}_s + \sigma(\vec{x}, t) \tag{4.4}$$

where $s(\vec{x}, t)$ is the entropy density and $\vec{J}_s$ is the flow density of the entropy $(s\dot{\vec{x}})$. We will use lower-case letters for the densities corresponding to the extensive variables denoted by the same upper-case letter. The quantity $\sigma(\vec{x}, t)$ is called the entropy production.

Besides the previous equation, it is also possible to write analogous continuity equations for the internal energy U and for the number of moles of the different chemical species present, N_i.

It is assumed that the fundamental Gibbs equation which relates the changes of the extensive variables in thermostatics,

$$TdS = dU - pdV + \sum_j \mu_j dN_j \tag{4.5}$$

(with T the absolute temperature, p the pressure and μ_k the chemical potential of the k-th chemical species) is valid locally, even for the non-equilibrium case, at least for systems which are not too far from equilibrium. This is equivalent to assuming that the functional dependence of S upon U, V and N_i is the same as in thermostatics, even though we are now dealing with functions which vary with time and space.

This is a kind of local equilibrium assumption, since the maintaining of the usual functional dependence is to be associated with the fact that, in a small portion of the system, conditions of slowly varying quasi-equilibrium are to be found. We again find ourselves in front of the idea of two clearly separated time scales, so that the macroscopic evolution of the system takes place over times which are long with respect to those necessary for the establishment of a quasi-Gibbs distribution in every small portion of the system. The hypothesis of local equilibrium also includes the idea that the spatial variation of the quantities considered is smooth.

Passing from the Gibbs equation to the densities and introducing the equations of balance for s, u, and n_j, after several steps which can be found in the standard textbooks on irreversible thermodynamics, we obtain:

$$\sigma = \vec{J}_q \cdot \vec{\nabla} \left(\frac{1}{T} \right) + \sum_i \vec{J}_i \cdot \vec{\nabla} \left(-\frac{\mu_i}{T} \right) + J_{ch} \frac{A}{T} \tag{4.6}$$

(with $\vec{J}_q$ the density of heat flow and $\vec{J}_i$ the density of material flow for the i-th component). The last term is associated with the possibility of chemical reactions occurring. Suppose we have

$$\nu_A A + \nu_B B + \ldots \rightleftarrows \nu_C C + \nu_D D + \ldots$$

then the variations dn_A, dn_B, etc. are in proportion, so that

$$dn_i = \nu_i d\xi$$

($\nu_i > 0$ for the products and $\nu_i < 0$ for the reagents). ξ is called the degree of advancement of the reaction. The quantity J_{ch} which appears in Eq. (4.6) is defined as:

$$J_{ch} = d\xi / dt$$

and is, therefore, in essence, the reaction rate. The affinity A is defined by the expression:

$$A = -\sum_j v_j \, \mu_j$$

and is the "driving force" of the reaction (which reaches equilibrium when A goes to zero).

Let us make a brief comment upon the form of Eq. (4.6) for the production of entropy. It takes the form of a sum of the products of the "flows" J times the corresponding generalised forces X:

	flow	conjugate force
matter	$\vec{J}_i$	$\vec{\nabla}(-\mu_i/T)$
heat	$\vec{J}_q$	$\vec{\nabla}(1/T)$
chemical	J_{ch}	A/T

Summarising in a more compact notation:

$$\sigma = \sum_k J_k X_k \tag{4.7}$$

Besides, every single flow, J_k, may depend upon all the generalised forces present:

$$J_k - J_k(X_1, \qquad X_{iv}) \tag{4.8}$$

The simplest case is that of linear irreversible thermodynamics, where it is assumed that such expressions are linear. In matrix notation:

$$J = LX \tag{4.9}$$

or rather:

$$X = RJ \tag{4.10}$$

where the matrix $R = L^{-1}$.

An important result, obtained by Onsager, is that under a wide range of general conditions, L is a symmetrical matrix:

$$L_{ik} = L_{ki} \tag{4.11}$$

The same is true for the inverse matrix R.

The Onsager equations (Eq. (4.11)) undoubtedly constitute the most famous result of linear irreversible thermodynamics, even though the range of their validity is still the subject of debate. These equations were employed to explain a series of symmetries which had been verified experimentally in different dissipative systems. This aspect of the theory is well-known and is exhaustively dealt with in the texts listed in the Bibliography.

We will limit ourselves, in this brief introduction of the most important definitions of irreversible thermodynamics, to give those of the Rayleigh dissipation function, ϕ, and the generating function, ψ:

$$\phi(J, J) = (1/2) \sum_{i,k} R_{ik} J_i J_k \tag{4.12}$$

$$\psi(X, X) = (1/2) \sum_{i,k} L_{ik} X_i X_k \tag{4.13}$$

These expressions allow us to write:

$$\sigma(J, X) = \sum_k J_k X_k = \sum_{k,m} L_{km} X_k X_m = 2\psi(X, X) \tag{4.14}$$

$$\sigma(J, X) = \sum_k J_k X_k = \sum_{k,m} R_{km} J_k J_m = 2\phi(J, J) \tag{4.15}$$

It follows that, in the linear range:

$$\sigma = 2\psi = 2\phi \tag{4.16}$$

and, obviously, $\phi = \psi$. Even though their numerical values are the same, the physical significance of the quantities introduced is different. $\phi(J, J)$ is, in fact, a function of the flows whereas $\psi(X, X)$ is a function of only the conjugate forces. $\sigma(J, X)$, however, is a combination of flows and forces. The importance of this distinction will become clearer in the following Sections.

4.2 THE LINEAR CASE

In order to understand the fundamental concepts of a subject it is often a good rule to study the simplest case, to avoid obscuring the key concepts beneath a formalism which, because of its generality, can often become very intricate. We will therefore consider the simple case of one macroscopic variable. Supposing that the system is not very far from equilibrium, we can assume an expression of the following type to be valid for the entropy:

$$s = -(c/2)\,\alpha^2 \tag{4.17}$$

where we have chosen α such that the equilibrium state corresponds to the value $\alpha = 0$.

If there is no entropy flow:

$$\sigma = \dot{s} = -c\,\alpha\,\dot{\alpha} \tag{4.18}$$

and the linear phenomenological relationship between the flow α and the conjugate force $-c\,\alpha$ is (see Eq. (4.10)):

$$R\dot{\alpha} = -c\,\alpha \tag{4.19}$$

By introducing $\gamma = c/R$, this becomes:

$$\dot{\alpha} = -\gamma\,\alpha \tag{4.20}$$

Onsager noted that this equation is completely analogous to the deterministic part of the Langevin equation, and suggested the addition to the r.h.s. of a delta-correlated fluctuating force:

$$d\alpha = -\gamma\alpha dt + dw \tag{4.21}$$

Up to here, the step from Eq. (4.20) to Eq. (4.21) is completely formal, and is suggested by analogy with Brownian motion, or more precisely by the discussion of the Langevin description of the Ehrenfest model (see Section 1.7). We will now discuss the plausibility of this operation.

First of all, α is a macroscopic variable. A macroscopic description is always a reduced description, in that one is limited to following the behaviour of only a few relevant variables, and does not follow the evolution of the numerous microscopic degrees of freedom. In the spirit of what we have accomplished so far, it is quite natural to simulate, to a first approximation, the effect of the removed degrees of freedom as a random perturbation of the deterministic equations which regulate the time evolution of the average values.

The hypothesis of additive white noise is the simplest quantitative translation of this intuition. For this to be realistic, we know that it is necessary to be able to identify two well-separated time scales in the system. Now, it is reasonable to suppose that the macroscopic flow results from the sum of many microscopic contributions which are essentially independent of one another, or almost. Therefore, to cause a perceivable change in α at the macroscopic level, it will be necessary to combine numerous microscopic events, and this would require times which are significantly longer than those typical of the microscopic events which give rise to the fluctuating force. The white noise hypothesis thus appears to be sufficiently well-founded to warrant a closer examination.

We are already familiar with Eq. (4.21) which was taken as a definition of the Ornstein-Uhlenbeck process in Section 1.5; the transition density is thus given by the following equation (see Eq. (1.57):

$$p(\alpha, t \mid \alpha_0) = \sqrt{\frac{\gamma}{2\pi D(1 - e^{-2\gamma t})}} \exp\left[-\frac{\gamma(\alpha - \alpha_0 e^{-\gamma t})^2}{2D(1 - e^{-2\gamma t})} \right] \tag{4.22}$$

where D is the diffusion coefficient of the fluctuating forces. At the limit $t \to \infty$, this expression becomes:

$$\lim_{t \to +\infty} p(\alpha, t \mid \alpha_0) = \sqrt{\frac{\gamma}{2\pi D}} \, e^{-\gamma \alpha^2 / 2D}$$

On the other hand, the Boltzmann principle assures us that the equilibrium distribution is proportional to $\exp[\Delta S/k]$, where ΔS is the entropy difference with respect to the equilibrium state (i.e. $- (c/2)\alpha^2$), and k is the Boltzmann constant. From this comparison we obtain the fluctuation-dissipation theorem:

$$D = k\gamma/c = k/R \tag{4.23}$$

where we have taken into account the previous definition of γ.

With this, we have arrived at the solution to our problem. From here onwards we will do nothing but transform the solution into a different form, for the following two reasons. Firstly, the new formulation will prove to be physically meaningful, it will enrich our understanding of the linear case and it will allow us to answer interesting questions. Secondly, it will constitute the basis for an extension of the theory to the case of nonlinear forces, for which we do not have an explicit solution such as Eq. (4.22).

4.3 PATH INTEGRALS

We have previously seen that the phenomenological equation (Eq. (4.20)) of linear irreversible thermodynamics may be interpreted, assuming the validity of the stochastic equation (Eq. (4.21)), as a time evolution law for the average values. In the linear case, therefore, the average values of the fluctuating variables obey the laws of the thermodynamics of irreversible processes. The same is true for the most probable values, which coincide with the average ones in the linear force case, where we deal with Gaussian processes.

The Ornstein-Uhlenbeck transition density, given by Eq. (4.22), is such that $p(\alpha, t \mid \alpha_0)\,d\alpha$ is the probability that a system which finds itself in α_0 at $t = 0$ will then be found near α at time t. The system, however, may go from α_0 to α in different ways. Suppose that we examine the system at an instant s between 0 and t. The corresponding value α_s may be different in different realizations of the process, because of the stochastic character of the fluctuating force $A(t)$.

This description can be made in more detail by introducing a continually increasing number of intermediate points. Let us divide the interval $[0, t]$ into $N + 1$ intervals of equal length:

$$t_0 = 0; \ t_1 = \epsilon ; \ldots t_j = j\epsilon ; \ldots t_{N+1} = t$$

Assuming that the extremes (t_0, α_0) and (t_{N+1}, α_{N+1}) are fixed, the probability of finding values around α_1 at time t_1, etc., is given by:

$$p(\alpha_{N+1}, t_{N+1} | \alpha_N, t_N) \, p(\alpha_N, t_N | \alpha_{N-1}, t_{N-1}) \cdots$$
$$\cdots p(\alpha_1, t_1 | \alpha_0, t_0) \, d\alpha_1 \ldots d\alpha_N \tag{4.24}$$

Let us now try to find an expression for the terms which appear in the r.h.s. of the preceding equation, i.e.:

$$p(\alpha_k + \delta\alpha_k, t_k + \epsilon | \alpha_k, t_k)$$

given that ϵ is small (later it will be assumed to tend towards 0).

Applying the Ornstein-Uhlenbeck formula with the approximation $\exp(-\gamma\epsilon) \sim 1 - \gamma\epsilon$, we obtain:

$$p(\alpha + \delta\alpha, t + \epsilon | \alpha, t) = \sqrt{\frac{\gamma}{2\pi D(2\gamma\epsilon)}} \, e^{-\frac{\gamma(\delta\alpha + \gamma\epsilon\alpha)^2}{2D(2\gamma\epsilon)}} =$$
$$= \sqrt{\frac{1}{4\pi D\epsilon}} \, e^{-\frac{1}{4D\epsilon}(\delta\alpha^2 + \gamma^2\alpha^2\epsilon^2 + 2\gamma\epsilon\alpha\delta\alpha)} \tag{4.25}$$

and the joint probability is, therefore:

$$p\begin{pmatrix} \alpha_{N+1} \cdots \alpha_0 \\ t_{N+1} \cdots t_0 \end{pmatrix} = \left(\frac{1}{4\pi D\epsilon}\right)^{\frac{N+1}{2}} \exp\Bigg\{ -\frac{1}{4D\epsilon} \cdot$$
$$\cdot \sum_{k=1}^{N+1} [(\alpha_k - \alpha_{k-1})^2 + \gamma^2\epsilon^2\alpha_{k-1}^2 + 2\gamma\epsilon\alpha_{k-1}(\alpha_k - \alpha_{k-1})] \Bigg\} \tag{4.26}$$

Recalling that

$$p(\alpha_{N+1}, t_{N+1} | \alpha_0, t_0) = \int p\begin{pmatrix} \alpha_{N+1} \cdots \alpha_0 \\ t_{N+1} \cdots t_0 \end{pmatrix} d\alpha_1 \ldots d\alpha_N$$

we thus obtain:

$$p(\alpha, t | \alpha_0) = \left(\frac{1}{4\pi D\epsilon}\right)^{\frac{N+1}{2}} \int d\alpha_1 \ldots d\alpha_N \exp\Bigg\{ -\frac{1}{4D} \cdot$$
$$\cdot \sum_{k=1}^{N+1} \left[\frac{(\alpha_k - \alpha_{k-1})^2}{\epsilon} + \gamma^2\epsilon\alpha_{k-1}^2 + 2\gamma\alpha_{k-1}(\alpha_k - \alpha_{k-1}) \right] \Bigg\} \tag{4.27}$$

with

$$\alpha_{N+1} = \alpha$$

We can try to make this expression a more exact one by allowing ϵ to tend towards zero and N towards infinity, so that the product $(N + 1)\epsilon$ remains constant. We then obtain:

$$p(\alpha, t \,|\, \alpha_0) = \lim_{\substack{N \to \infty \\ \epsilon \to 0 \\ (N+1)\epsilon = \text{const.}}} \left(\frac{1}{4\pi D\epsilon}\right)^{\frac{N+1}{2}} \int d\alpha_1 \ldots d\alpha_N \exp\left\{-\frac{1}{4D} \int_0^t (\dot{\alpha}^2 + \right.$$

$$\left. + \gamma^2 \alpha^2 + 2\gamma\alpha\dot{\alpha}) \; ds \right\} \tag{4.28}$$

Richard Feynman, while studying quantum mechanical problems, considered expressions of the same type, and coined the expression"path integrals". The deep relationship existing between quantum mechanics and the theory of stochastic processes is a fascinating subject on which we will not dwell, as it is outside the scope of the present work. Eq. (4.28) can be interpreted by realizing that the transition density between the point $(0, \alpha_0)$ and the point (t, α) is the sum (i.e. the integral), over all the possible paths between these two points, of the probability density of every single path in (t, α) space (see Fig. 4.1).

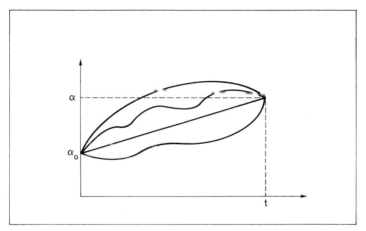

Fig. 4.1. - Paths in (t, α) space between two fixed extremes.

We can adopt a new symbolism, rewriting Eq. (4.28) as:

$$p(\alpha, t \,|\, \alpha_0) = \int_0^t \mathscr{D}\alpha(s) \, p([\alpha(s)]) \tag{4.29}$$

where:

$$p([\alpha(s)]) = \exp\left\{-\frac{1}{4D} \int_0^t (\dot{\alpha}^2 + \gamma^2\alpha^2 + 2\gamma\alpha\dot{\alpha}) \, ds \right\}$$

Note the use of the square brackets: $p([\alpha(s)])$ indicates the probability density for a particular path $[\alpha(s)]$. We recall that the probability for a single path vanishes, for the same reasons that render the probability for a particular numerical value of a stochastic variable with continuous range equal to zero. In this latter case, as we know, we introduce a probability density $p(x)$, so that $p(x)\,dx$ is the probability to obtain values between x and $x + dx$. In an analogous way we consider now a path probability density $p([\alpha(s)])$ such that $p([\alpha(s)]) \, \mathscr{D}\alpha(s)$ equals the probability that the system evolves in an "infinitesimal tube" around $[\alpha(s)]$.

It is important to underline that Eq. (4.29) is only a different, more physically illuminating form of Eq. (4.28), and does not contain any additional information. The normalising factor is taken account of in the "measure in path space", denoted by $\mathscr{D}\alpha(s)$.

It is worth pausing for a moment upon the concept of path probability. At the limit for $N \rightarrow \rightarrow \infty$, the sequence $\alpha_0, \ldots \alpha_{N+1}$ becomes a path $[\alpha(s)]$. Because of the fluctuating character of the force $A(t)$, the time evolution described by the Langevin equation (Eq. (4.21)) is very irregular and, in the white noise approximation, it can not even be drawn. The normal recipe for associating an experimental value to a stochastic equation is that of considering an enormous number, M, of mental copies of the system, all prepared in the same way from the macroscopic point of view, and to consider their average. In this way, at every point in time, we will have:

$$\langle \alpha(t) \rangle = \lim_{M \rightarrow \infty} \frac{1}{M} \sum_{k=1}^{M} \alpha_k(t)$$

The time evolution law for the average values is given by the phenomenological equation (Eq. (4.20)), so that:

$$\langle \alpha(t) \rangle = \alpha_0 e^{-\gamma t} \tag{4.30}$$

Therefore, even if each single realization of the stochastic process is irregular, and corresponding trajectories can not be drawn, the average values show a regular behaviour over time which is coherent with experimental observations. Knowing, however, that it is possible to observe macroscopic deviations from the average behaviour, it is reasonable to ask what the probability is of a path which is regular but different from that described by Eq. (4.30). For this reason it is interesting to consider the probability density $p[\alpha(s)]$, with $\alpha(s)$ as a regular function. The path probability of Eq. (4.29) can be rewritten as:

$$p([\alpha(s)]) = \exp\left\{ -\frac{1}{4D} \int_0^t (\dot{\alpha} + \gamma\alpha)^2 \, ds \right\} \tag{4.31}$$

This reaches a maximum when the integral in the exponent is at a minimum or, as the integrand is not negative, when the phenomenological equation (Eq. (4.20)) is satisfied, a condition that renders it exactly zero.

In conclusion, the most probable path is that described by the laws of irreversible thermodynamics. Since we have a Gaussian case, the most probable path coincides with the average one, so that we have obtained an already known result.

Let us now present a different way of deriving an expression for the most probable path: although it might seem superfluous at this point, it will be of help in the non linear case where no explicit solution, analogous to the Ornstein-Uhlenbeck one, is available. In order to determine the most probable path with fixed initial conditions, let us remark that it will be such as to maximise the transition probability density at each infinitesimal step. From Eq. (4.25) we obtain:

$$\delta\alpha^2 + 2\gamma\epsilon\alpha \, \delta\alpha = minimum \tag{4.32}$$

Deriving with respect to the increment yields:

$$\frac{\delta\alpha}{\epsilon} = -\gamma\alpha$$

and finally

$$\dot{\alpha} = -\gamma\alpha \tag{4.33}$$

Deriving with respect to $\delta\alpha$ is therefore equivalent to deriving with respect to the final point, and also, since ϵ is fixed, with respect to $\delta\alpha/\epsilon$.

We will now show that it is possible to express the path probability in terms of the thermodynamic quantities σ, ϕ and ψ.

In our case, these are:

$$\sigma = \dot{s} = -c\alpha\dot{\alpha}$$

$$\phi = \frac{1}{2} R \,\dot{\alpha}^2 \tag{4.34}$$

$$\psi = \frac{1}{2R} X^2 = \frac{c^2}{2R} \alpha^2$$

The expression already obtained for the path probability density can then be rewritten in the form:

$$p([\alpha(s)]) = \exp\left\{-\frac{1}{2k}\int_0^t (\phi + \psi - \sigma)\,d\tau\right\} =$$

$$= \exp\left\{\frac{1}{2k}\left[S(\alpha) - S(\alpha_0) - \int_0^t (\phi + \psi)\,d\tau\right]\right\} \tag{4.35}$$

The deterministic path is therefore described by:

$$\int_0^t (\phi + \psi - \sigma)\,d\tau = \text{minimum}$$

$$\sigma = \phi + \psi \tag{4.36}$$

Finally, in the literature, a certain amount of confusion exists with regard to the "path probability". This is due to the introduction of variational principles which, by analogy with its use in other areas of physics, should provide an elegant synthesis of the theoretical foundations. Whilst in mechanics these principles may be postulated (the Hamilton principle), a similar operation is not allowed in our case. In fact, once we accept the logic of associating the deterministic equation (Eq. (4.20)) with its stochastic analogue (Eq. (4.21)), the problem is completely defined, and every statistical property of the process has to be compatible with the Ornstein-Uhlenbeck formula.

In particular, Onsager considered the problem of maximising the probability of a certain path between two fixed extremes, that is, the problem of minimising

$$\int_0^t (\dot{\alpha}^2 + \gamma^2\alpha^2 + 2\gamma\alpha\dot{\alpha})\,d\tau = \text{minimum}$$

subject to the conditions

$$\alpha(0) = \alpha_0 \qquad \alpha(t) = \alpha$$

The solution, as is well-known, is given by the Euler-Lagrange equation:

$$\ddot{\alpha} = \gamma^2\alpha$$

the general solution of which is:

$$\alpha(s) = A\, e^{-\gamma s} + B\, e^{\gamma s}$$

This is the path of maximum probability between the two fixed extremes $(0, \alpha_0)$ and (t, α). There is no relationship between this and the most probable path starting from a given initial condition but with an unknown second extreme. The latter case corresponds to that defined by the phenomenological equation of irreversible thermodynamics. In fact, the second extreme (t, α) may have an extremely low probability of being reached, so that the most probable path, amongst those that reach it, may be very different from the most probable path amongst those which pass through the point $(0, \alpha_0)$.

4.4 THE NONLINEAR CASE

Let us now see what happens in the case where the phenomenological equation is no longer required to take the linear form of Eq. (4.19) or (4.20), but is of the type:

$$\dot{\alpha} = (1/R)\, f(\alpha) \tag{4.37}$$

where f may be a non-linear function of its argument. On the basis of the same considerations as in the linear case, we introduce the stochastic differential equation:

$$\dot{\alpha} = (1/R)\, f(\alpha) + A(t) \tag{4.38}$$

where $A(t)$ is a white fluctuating force, with a diffusion coefficient D. Let us also suppose that the generalised force is the derivative of the entropy, in agreement with the general scheme of the thermodynamics of irreversible processes:

$$f(\alpha) = \frac{d}{d\alpha}\, S(\alpha) \tag{4.39}$$

It is not possible, given the non-linearity of the force, to obtain an explicit expression for the transition density. However, a representation can be given in the form of path integrals. It would be possible to formally deduce this expression (Girsanov's theorem) but, continuing in the spirit of this book, we prefer to justify it with more simple considerations of plausibility.

We therefore return to the linear case for our inspiration. By putting

$$f(\alpha) = -c\,\alpha \tag{4.40}$$

it is possible to directly verify that Eq. (4.29) for the path probability can be written in the form:

$$p([\alpha(s)]) = \rho([\alpha(s)])\, p_f([\alpha(s)]) \tag{4.41}$$

where:

$$p_f([\alpha(s)]) = \exp\left[-\frac{1}{4D} \int_0^t \dot{\alpha}^2\, ds\right]$$

$$\rho([\alpha(s)]) = \exp\left[\frac{1}{2D} \int_{\alpha_0}^{\alpha} \frac{1}{R} f(\alpha')\, d\alpha' - \frac{1}{4D} \int_0^t \left(\frac{1}{R} f(\alpha(s))\right)^2 ds\right]$$

where use has been made of the relationship

$$\dot{\alpha}\, dt = d\alpha$$

The physical interpretation of the distinction between p_f and ρ is that the first represents the probability density for a free path (Wiener process), that is, in the case of zero force, as can be verified by putting $\gamma = 0$ in Eq. (4.29). ρ is then a path functional which transforms the density of the Wiener process into the driven process which has the same diffusion coefficient.

In this way we have eliminated every explicit reference to the hypothesis of linearity, and we can ask ourselves whether or not we have an acceptable expression for the general case.

First of all, however, we must clarify whether the stochastic integral which appears in the expression for ρ, $\int f(\alpha)\,d\alpha$, is to be interpreted as an Itô or a Stratonovich integral. To this end, we note that the infinitesimal transition density can be obtained directly from the representation in terms of path integrals, since in the case of the infinitesimal step all the integrations over the intermediate points vanish. We can thus compare the results obtained according to both Stratonovich and Itô with the Ornstein-Uhlenbeck infinitesimal transition density for the case of a linear force (Eq. (4.25)). If we choose Stratonovich,

$$\int_{\alpha_0}^{\alpha} f(\alpha')\,d\alpha' = S(\alpha) - S(\alpha_0)$$

and the infinitesimal density is ($M(\epsilon)$ is a normalisation factor):

$$p(\alpha + \delta\alpha,\, t + \epsilon\,|\,\alpha,\, t) = \frac{1}{M(\epsilon)}\,\exp\left\{\frac{1}{2DR}\left(-\frac{c}{2}(\alpha + \delta\alpha)^2 + \right.\right.$$

$$\left.\left. +\frac{c}{2}\alpha^2\right) - \frac{1}{4D}\,\epsilon\left(\frac{\delta\alpha^2}{\epsilon^2} + \frac{c^2\alpha^2}{R^2}\right)\right\} = \tag{4.42}$$

$$= \frac{1}{M(\epsilon)}\,\exp\left\{-\frac{1}{4D}\left[\gamma\delta\alpha^2 + 2\gamma\alpha\delta\alpha + \frac{\delta\alpha^2}{\epsilon} + \gamma^2\alpha^2\epsilon\right]\right\}$$

which differs from the correct expression (Eq. (4.25)) due to the term $\gamma\delta\alpha^2$. Interpreting the stochastic integral according to Itô's recipe, however, i.e.

$$\int_{\alpha_0}^{\alpha} f(\alpha')\,d\alpha' = S(\alpha) - S(\alpha_0) - D\int_0^t f'(\alpha(s))\,ds \tag{4.43}$$

we obtain:

$$p(\alpha + \delta\alpha,\, t + \epsilon\,|\,\alpha,\, t) = \frac{1}{M(\epsilon)}\,\exp\left\{-\frac{1}{4D}\left[2\gamma\alpha\delta\alpha + \right.\right.$$

$$\left.\left. +\frac{\delta\alpha^2}{\epsilon} + \gamma^2\alpha^2\epsilon + (\gamma\delta\alpha^2 - 2\gamma D\epsilon)\right]\right\} \tag{4.44}$$

This expression is equal to the Ornstein-Uhlenbeck expression since the term $\gamma\delta\alpha^2$ is cancelled by Itô's corrective term. We recall, in fact, that $\delta\alpha^2 = 2D\,\epsilon$, so that the contents of the last bracket on the r.h.s. is zero.

Besides, from the previous equation we obtain the correct expression for the normalisation term:

$$M(\epsilon) = \sqrt{4\pi D\,\epsilon} \tag{4.45}$$

The interpretation according to Itô is therefore correct for the linear case, and we formulate the hypothesis that it is generally valid. The expression for the transition density (Eq. (4.41)) can then be rewritten as:

$$p([\alpha(s)]) = \exp\left[\frac{S(\alpha) - S(\alpha_0)}{2k}\right] K(\alpha, t \mid \alpha_0)$$

$$K(\alpha, t \mid \alpha_0) = \exp\left[-(1/2) \int_0^t \mathscr{L}(\alpha(s), \dot\alpha(s)) \, ds\right]$$

$$\mathscr{L}(\alpha, \dot\alpha) = \frac{\dot\alpha^2}{2D} + V(\alpha)$$

$$V(\alpha) = (1/R) [f'(\alpha) + (1/2k) f^2(\alpha)]$$

(4.46)

In this expression $K(\alpha, t \mid \alpha_0)$ is the kernel or propagator, $\mathscr{L}$ the thermodynamic Lagrangian, and $V(\alpha)$ is called the Onsager-Machlup potential. We will comment later upon the meaning of the various terms defined here.

We will now provide a very plausible argument that the previous equation is also correct for the nonlinear case. We will show that from Eq. (4.46) it is possible to obtain a differential equation for the transition density, and that this corresponds to the Fokker-Planck equation.

In order to simplify the notation, we will drop the indication of the initial point, putting:

$$\phi(\alpha, t) = K(\alpha, t \mid \alpha_0)$$

(4.47)

From the Chapman-Kolmogorov equation which, for the sake of simplicity, we write as

$$p(3 \mid 1) = \int (d2) \, p(3 \mid 2) \, p(2 \mid 1)$$

it follows that:

$$p(\alpha, t + \epsilon \mid \alpha_0, t_0) = \int_{-\infty}^{+\infty} d\eta \, p(\alpha, t + \epsilon \mid \alpha + \eta, t) \cdot p(\alpha + \eta, t \mid \alpha_0, t_0)$$

Passing to the corresponding equation for the propagator:

$$K(\alpha, t + \epsilon \mid \alpha_0, t_0) = \frac{1}{M(\epsilon)} \int_{-\infty}^{+\infty} d\eta \, e^{-\mathscr{L}(\alpha, \eta/\epsilon)} K(\alpha + \eta, t \mid \alpha_0, t_0)$$

We now introduce the definition of ϕ (Eq. (4.47)) and develop the various terms up to the first order in the time increment ϵ and to the second order in the increment $\eta = \Delta\alpha$. We therefore have:

$$\phi(\alpha, t + \epsilon) = \frac{1}{M(\epsilon)} \int_{-\infty}^{+\infty} d\eta \exp\left\{-\frac{\eta^2}{4D\epsilon}\right\} \left\{\left(1 - \frac{\epsilon}{2} V(\alpha)\right) \cdot \right.$$

$$\cdot \left[\phi(\alpha, t) + \eta \frac{\partial}{\partial \alpha} \phi(\alpha, t) + \frac{1}{2} \eta^2 \frac{\partial^2}{\partial \alpha^2} \phi(\alpha, t)\right]$$

The contribution from the linear terms in η vanishes on the grounds of symmetry, whereas the remaining integrals are Gaussian. Thus:

$$\phi(\alpha,\ t+\epsilon) = \frac{1}{M(\epsilon)} \left\{ \sqrt{4\pi D\epsilon} \left(1 - \frac{1}{2}\ \epsilon V(\alpha)\right)\phi(\alpha,\ t) + \right.$$

$$\left. + D\ \epsilon \sqrt{4\pi D\epsilon}\ \frac{\partial^2}{\partial\alpha^2}\ \phi(\alpha,\ t) \right\} \tag{4.48}$$

Taking the limit $\epsilon \to 0$, we can determine the normalisation coefficient:

$$M(\epsilon) = \sqrt{4\pi D\ \epsilon}$$

(as in the linear case, Eq. (4.45)). Eq. (4.48) thus becomes:

$$\phi(\alpha,\ t+\epsilon) - \phi(\alpha,\ t) = \epsilon \left[-\frac{1}{2}\ V(\alpha)\ \phi(\alpha,\ t) + D\ \frac{\partial^2}{\partial\alpha^2}\ \phi(\alpha,\ t) \right] \tag{4.49}$$

Dividing both sides by ϵ and taking the limit $\epsilon \to 0$, we obtain a differential equation for the propagator K:

$$\frac{\partial}{\partial t} K(\alpha,\ t\,|\,\alpha_0) = \left[D\ \frac{\partial^2}{\partial\alpha^2} - (1/2)\ V(\alpha) \right] K(\alpha,\ t\,|\,\alpha_0) \tag{4.50}$$

It can be verified that this corresponds to the equation derived from the Fokker-Planck equation:

$$\frac{\partial}{\partial t} p(\alpha,\ t\,|\,\alpha_0) + \frac{1}{R}\ \frac{\partial}{\partial\alpha}\ (f(\alpha)\ p) = D\ \frac{\partial^2 p}{\partial\alpha^2}$$

by introducing the definition:

$$p(\alpha,\ t\,|\,\alpha_0) = \exp\left\{ \frac{S(\alpha) - S(\alpha_0)}{2\,k} \right\} K(\alpha,\ t\,|\,\alpha_0)$$

To summarise, we have:
- studied the linear case, demonstrating that the path probability can be expressed as a function of the thermodynamic variables σ, ϕ and ψ;
- rewritten the expression for the path probability density in a form which makes it a good candidate for a generalisation to the nonlinear case;
- verified that there is a good possibility of it being a suitable candidate because it provides the correct time evolution equation for the propagator.

As we have already mentioned, it would be possible to provide a more formal demonstration of the validity of this representation. But we followed a simpler route which also seems more attractive from a physical point of view. Incidentally, we have employed the same technique with which Feynman demonstrated that the path integral representation of the quantum propagator, suggested by him, led to the Schrödinger equation.

Let us now comment upon the significance of the introduction of the propagator K, of the Lagrangian $\mathcal{L}$ and of the Onsager-Machlup potential V. For the propagator, the first Eq. (4.46) shows that the transition density can be expressed as the product of two terms, the first of which depends only upon the entropy difference between the initial and final points, whereas the second is itself the propagator. The thermodynamic Lagrangian and the Onsager-Machlup potential in the expression for K play formally analogous roles to the normal Lagrange function and the potential in the case of the relationship between classical and quantum mechanics.

It is interesting to introduce here a new quantity, the joint entropy, σ_j, defined as:

$$K(\alpha, t \,|\, \alpha_0) = \exp\left[(1/2k)\,\sigma_j(\alpha, t \,|\, \alpha_0)\right] \tag{4.51}$$

The transition probability can thus be rewritten as:

$$p(\alpha, t \,|\, \alpha_0) = \exp\left[(1/2k)\,(S(\alpha) - S(\alpha_0) + \sigma_j(\alpha, t \,|\, \alpha_0))\right] \tag{4.52}$$

The Boltzmann principle assures us that

$$\lim_{t \to +\infty} p(\alpha, t \,|\, \alpha_0) = \exp\left[S(\alpha)/k\right] \tag{4.53}$$

Therefore:

$$\lim_{t \to +\infty} \sigma_j(\alpha, t \,|\, \alpha_0) = S(\alpha) + S(\alpha_0) = \sigma_{j\infty}(\alpha, \alpha_0) \tag{4.54}$$

This explains why σ_j has been called the joint entropy. It becomes the sum of the entropies of the initial and final states in the limit of infinitely long times, whereas for finite times it takes into account the statistical correlations between the two states.

Introducing the transformation (Eq. (4.51)) in the time evolution equation for the kernel (Eq. (4.50)), we obtain the time evolution equation for the joint entropy:

$$\frac{\partial}{\partial t}\,\sigma_j(\alpha, t \,|\, \alpha_0) = D\,\frac{\partial^2}{\partial \alpha^2}\,\sigma_j + \frac{D}{2k}\left(\frac{\partial \sigma_j}{\partial \alpha}\right)^2 - k\,V(\alpha) \tag{4.55}$$

It can be directly verified that the asymptotic expression $\sigma_{j\infty}$ (Eq. (4.54)) is effectively a stationary solution of this equation.

The concept of joint entropy is therefore consistent with the Boltzmann principle, and Eq. (4.52) may be quite justly called the "kinetic analogue of the Boltzmann principle". It does, in fact, provide the transition density between two states as a function of the difference between their entropies and of their joint entropy, as the Boltzmann principle provides the probability density of a state as a function of its entropy.

We conclude our discussion with some remarks about the most probable path. The infinitesimal transition probability density can be expressed as follows:

$$p(\alpha + \delta\alpha, t + \epsilon \,|\, \alpha, t) = \frac{1}{M(\epsilon)}\,\exp\left\{\frac{S(\alpha + \delta\alpha) - S(\alpha)}{2k} - \right.$$

$$\left. - \frac{\epsilon}{2}\left[\frac{1}{2D}\left(\frac{\delta\alpha}{\epsilon}\right)^2 + \frac{1}{R}\left(f'(\alpha) + \frac{1}{2k}\,f^2(\alpha)\right)\right]\right\}$$

It attains its maximum value when

$$\frac{1}{2k}\left(f(\alpha)\,\delta\alpha + \frac{1}{2}\,f'(\alpha)\,\delta\alpha^2\right) - \frac{\delta\alpha^2}{4D\epsilon} - \frac{\epsilon}{2R}\,f'(\alpha) - \frac{\epsilon}{4kR}\,f^2(\alpha) = \max$$

Differentiating with respect to $\delta\alpha$, gives:

$$\frac{\delta\alpha}{\epsilon} = \frac{1}{R}\,f(\alpha) + \frac{1}{2k}\,f'(\alpha)\,\delta\alpha \ .$$

A comparison of the $\delta\alpha \to 0$ limit of this expression with Eq. (4.37), which describes the "deterministic path" associated with the stochastic process defined by the nonlinear Langevin equation (Eq. (4.38)), illustrates the central role of the deterministic path in the set of random paths. It is also to be noted that it would be an oversimplification to say that the deterministic path is the most probable amongst all paths. For example, in the case of a double symmetric potential well, with the initial condition corresponding to the relative maximum of the potential separating the two wells, the deterministic path, given by the solution of Eq. (4.37), describes a static situation in which the point remains indefinitely in the equilibrium position, whereas the most probable path leads the system towards the potential minima. Finally, it is to be noted that, contrary to the linear case, in the case of nonlinear Langevin equations the deterministic path is not coincident with the average path. In fact, using Eq. (4.38)), we obtain

$$\frac{\partial}{\partial t} \langle \alpha \rangle = \langle f(\alpha) \rangle \tag{4.56}$$

The solution to this equation, which gives the average path, is different from that of Eq. (4.37), which defines the deterministic path, since normally, in the nonlinear case, $f(\langle \alpha \rangle) \neq \langle f(\alpha) \rangle$.

5

Stability and Bifurcations in Dynamical Systems

5.1 INTRODUCTION

Up to now, we have examined the effects of fluctuations on dynamical systems. From the case of Brownian motion, where the differential equation contains simply a linear term due to the dissipation, we went on to study slightly more complicated systems. For the latter, we introduced the concept of stable states, examining the effect of fluctuations on the stability and the possibility of transitions induced by fluctuations between different equilibrium states.

Our considerations, at first relating to mechanical systems, were later extended to any type of system. The variables considered, therefore, can include not only coordinates but also concentrations, electric charges, field intensities and other physical variables. In general, we call these variables "order parameters" because they are often aggregated variables which describe the collective behaviour of many systems which are similar or identical (atoms, molecules, single individuals of a population).

The analysis carried out in Chapter 4 showed how, in the stochastic process

$$\dot{\alpha} = f(\alpha) + A(t) \tag{5.1}$$

a privileged position is held by the deterministic path, which, by definition, is the solution of the deterministic equation:

$$\dot{\alpha} = f(\alpha) \tag{5.2}$$

The physical origins of this "centrality" of the deterministic path can be understood by considering the symmetry of the Langevin fluctuating force around the origin. The study of deterministic dynamical systems of the above type is thus important. Also, it can be seen that the order parameters discussed here are often average values, resulting from the sum of an extremely large number of elementary contributions. In such cases, under rather general hypotheses, the relative amplitude of the fluctuations is scaled by using the reciprocal of the square root of N, the number of elementary contributions, that is,

$$\sqrt{\frac{\langle (x - \langle x \rangle)^2 \rangle}{\langle x \rangle^2}} \approx \sqrt{\frac{1}{N}}$$

Thus, if N is very large, the amplitude of the fluctuations is very small and often can not be observed experimentally.

Therefore, in these cases a deterministic analysis is sufficient for describing the behaviour of the system, with one obvious precaution: we must not forget the existence of fluctuations which render the unstable states physically impossible, such as those, for example, corresponding to relative maxima in the potential.

A further argument in favour of a more detailed study of deterministic systems is as follows. Very often the main interest in an analysis is the asymptotic behaviour of a system rather than the transients. The stationary solution of the Fokker-Planck equation associated with the Langevin equation (5.1) is

$$p(\alpha) = \exp\left[- U(\alpha)/D\right]$$

where $U(\alpha)$ is the potential from which the generalised force $f(\alpha)$ can be derived

$$f(\alpha) = -\frac{dU(\alpha)}{d\alpha}$$

and D denotes the diffusion coefficient of the fluctuating force. Thus an analysis of the potential is sufficient for a gross characterisation of the asymptotic probability distribution. This result may also be described by saying that the attractors of the deterministic system correspond to local maxima of $p(\alpha)$, when long time periods are considered.

We can thus summarise the logical process linking the three descriptive levels of the physical world as follows. Starting from a microscopic description and eliminating the fast degrees of freedom, a reduced mesoscopic description is obtained which contains noise terms. In the case of linear interaction between the degrees of freedom of the system, these noise terms, at a suitable level of approximation, are represented by additive Langevin forces. The latter then appear in the dynamical equations without ever being multiplied by a function of the order parameters (see Chapter 3).

Thus, the arguments presented in Chapter 4, as well as those immediately above, are valid and we can then pass from a mesoscopic (stochastic) to a macroscopic (deterministic) description as long as the relative amplitude of the fluctuations is small.

The situation will change, however, if the interactions between the degrees of freedom are nonlinear. In this case adiabatic elimination will generally give rise to mesoscopic equations for the order parameters containing multiplicative noise terms, in which the fluctuating force is multiplied by a function of the order parameters. Terms of this type may also arise from a coupling of the system with a surrounding fluctuating environment. As we have already shown in Chapter 3, these terms may give rise to "counterintuitive" effects, known as "noise-induced transitions" (the transition shown in Fig. 3.2, for example), which do not have a deterministic counterpart. Another example of the effects of multiplicative noise is given in Section 6.8.

In these cases the analysis of the deterministic model, in which the fluctuating forces are assigned their exact average value, is no longer sufficient, since it is no longer capable of reproducing the phenomenology of the corresponding stochastic model. Even in this case, however, a deterministic analysis may be useful, as long as it is interpreted with the necessary care. In fact, noise-induced transitions are observed only when the intensity of the multiplicative noise exceeds high threshold values. Where this is not the case, a deterministic study is useful for understanding at least some features of the dynamic behaviour of the model.

We will now examine the effect of varying some of the parameters on the stability properties of physical systems. This corresponds to a situation which frequently occurs in the laboratory, where it is possible to externally control the values of several quantities (thus called "control parameters") which influence the system under examination. As we shall see, in many physical systems a simple variation in an external parameter (temperature, light stimulation, applied electric field, feed rate of reactants in a chemical reactor, etc.) may induce (with the help of fluctuations) a transition from the intial state (rendered unstable) towards new stable states possibly originating from the same variation in the external parameter.

In many physical systems (for example, lasers), this transition involves passing from a more "disordered" state to a more "ordered" one. In these cases we will thus talk of "organisation", or better still "self-organisation", given that the external stimulus (the variation of a parameter) does not, in itself, bring about the change. It is rather the internal dynamics of the system (with the interaction between microscopic variables and the emergence of several order parameters) which gives rise to an organised state.

Here we will only examine relatively simple models of organisation, which can be described by means of attractors (static or dynamical) for several order parameters. A more complete analysis of organisation within a space-time framework would, in fact, require the use of partial differential equations, but this lies beyond the scope of this book.

In any case, for a more complete understanding of these questions, even in the simplest systems, a study of stability in deterministic dynamical systems is necessary.

5.2 STABILITY OF THE SOLUTIONS OF A FIRST-ORDER DIFFERENTIAL EQUATION

Let us take the deterministic part of a Langevin equation, which we will write as:

$$\dot{x} = -K x \tag{5.3}$$

where K is a positive and time-independent constant. The solution of this equation is simply:

$$x = x_0 \exp[-K t] \tag{5.4}$$

and the origin is a point of "stable equilibrium". The use of this definition expresses the fact that if the system, initially in equilibrium, undergoes a sufficiently small perturbation, it remains, for any given period of time, within a limited neighbourhood of the point of equilibrium. In the case we are considering, the point of equilibrium is also "asymptotically stable", because the extent of the neighbourhood in which the representative coordinate of the system is to be found tends to zero with increasing time. The point of equilibrium is unstable when $K < 0$.

The previous linear equation is a particular case of the more general equation

$$\dot{x} = F(x) \tag{5.5}$$

In the case where the variable x represents a position, the latter equation, in its turn, may be considered as an approximate expression of the second law of dynamics in the presence of very high damping.

Similar equations are often encountered in the thermodynamics of irreversible processes (see Chapter 4) and in this case the term $F(x)$ is often called a "generalised force".

If an explicit form of the potential function $V(x)$, such that $F(x) = -dV(x)/dx$, can be found, then a study of the stability may be simply carried out by examining the shape of the potential, and in particular its second derivative with respect to x at the point of equilibrium. If the second derivative of the potential is positive (and therefore the infinitesimal amount of "work" carried out by the force F on moving away from the point of equilibrium is negative) then the equilibrium is stable.

Let us now turn to the case of a nonlinear "force" of the type

$$F(x) = -Kx - K_1 x^3 \qquad (K_1 > 0) \tag{5.6}$$

which corresponds to the potential

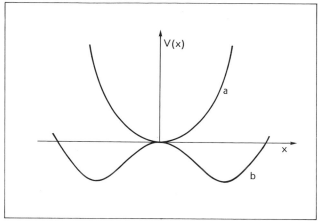

Fig. 5.1 - Qualitative behaviour of the potential $V(x) = (K/2)x^2 + (K_1/4)x^4$
(with $K_1 > 0$) corresponding to the cases $K > 0$ and $K < 0$ (curves
(a) and (b), respectively).

$$V(x) = \frac{K}{2}x^2 + \frac{K_1}{4}x^4 \tag{5.7}$$

From here on, we will suppose that $K_1 > 0$. If the coefficient K is also positive, as in the case of the deterministic part of the Langevin equation (5.3), the nonlinear term does not substantially modify the stability characteristics of the system, which maintains the origin as a point of stable equilibrium (Fig. 5.1a). If, however, $K < 0$ (with $K_1 > 0$) the potential takes on the shape shown in Fig. 5.1b. Here, besides the unstable equilibrium state corresponding to $x = 0$, there are two additional stable points of equilibrium, due to the presence of the nonlinear term, which are symmetrical about the origin and have the coordinates

$$x_0^{(1,2)} = \pm\,[\,|\,K\,|\,/K_1\,]^{1/2} \tag{5.8}$$

In certain cases only one of these points may have a physical meaning, but this does not detract from the generality of our considerations.

Let us now suppose that in the simple nonlinear equation

$$\dot{x} = -Kx - K_1 x^3 \tag{5.9}$$

K_1 remains constant and positive whereas K is allowed to vary passing from positive to negative values. As soon as K passes through zero, the previously stable point of equilibrium (the origin) becomes unstable and there is a "bifurcation" of the stable solution into two stable and one unstable solutions, according to the scheme

$$stable \begin{cases} stable \\ unstable \\ stable \end{cases}$$

The system, however, may be found only in one or the other of the two possible positions of stable equilibrium, so that the emergence of the instability at the origin gives rise to "symmetry breaking".

To better understand the effect of this "symmetry breaking" due to the variation of the parameter K, we suppose first of all that, for $K > 0$, the system finds itself at $x = 0$, and then we modify the value of K (which, we assume, can be controlled externally) until it becomes slightly negative. In the absence of fluctuations, the system would remain in the position of equilibrium corresponding to $x = 0$, which has now become unstable, even for negative values of K. The stability or instability of an equilibrium position, in fact, take on their full physical significance only in the presence of fluctuations which can "explore" neighbouring states around the state of equilibrium. On this point, we recall the definition of stability given above.

In other words, it is the presence of fluctuations which triggers off the transition of the system from the previous situation of stable equilibrium (now unstable) towards the new equilibrium state.

Far away from the bifurcation point, it is the deterministic equation which governs the behaviour of the system, in the sense that the presence of fluctuations simply determines a certain probability distribution about the average value of the "order parameter". In the vicinity of the bifurcation, however, the fluctuations decide towards which situation the system will evolve.

In the case considered so far, the coordinate of the new equilibrium position in which the system finds itself is a continuous function of the parameter K. Any function of state of the system will thus be continuously dependent upon the parameter K. Corresponding to the variation in the control parameter K, we thus have a "continuous transition" of the type shown in Fig. 5.2. This type of bifurcation is called a "pitchfork" bifurcation, after the shape of the graphical representation of the equilibrium values of the variable x. That is, from a single equilibrium value, three different ones are obtained when the parameter K passes through zero. We will later discuss other types of bifurcations corresponding to continuous transitions.

We will now consider a simple example in which "discontinuous transitions" are possible, that is, transitions which involve sharp changes in the values of the representative variables of the system following even small variations in the control parameter. To streamline the demonstration, we will refer to a particular physical meaning of the equations, without wishing to limit the general validity of the results which will be presented.

Let us thus consider a dynamical model for the growth of a "population" and restrict our considerations to the one-dimensional case, for simplicity. In this context, the word "population" takes on a completely general meaning, and is not necessarily intended in a strictly biological sense. As we shall see in the Chapter 6, even physical entities such as coherent photons in a laser cavity or the different chemical compounds in a complex reaction may be considered as "populations", in the sense that they obey equations of the type developed below. It is also to

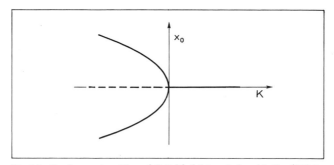

Fig. 5.2 - Continuous transition of "pitchfork" type induced in Eq. (5.9) by varying the control parameter K. The stable equilibrium solutions (x_0) lie on the solid branches whereas the dashed one represents the unstable solutions.

be noted that the possibility of describing order parameters of complex systems in terms of "population dynamics" seems to be one of their most general characteristics, which further justifies the choice of this particular example.

Let us start from the simple model

$$dx/dt = bx - mx \qquad (5.10)$$

where x is the number of individuals in the "population" considered, b is the birth rate and m the death rate. If $b > m$ the origin is clearly a point of unstable equilibrium, whereas it is stable if $b < m$.

In the former case, in order to avoid a population "explosion", it is physically meaningful to introduce a term which limits the growth due to the scarcity of resources. Such a limitation may be represented, for example, in a very simple way by using a quadratic "saturation" term with negative sign, so that the equation becomes

$$dx/dt = bx \, (1 - x/A) - mx \qquad (5.11)$$

Besides $x = 0$, there is now a second equilibrium solution $x_0 = A(1 - m/b)$, which is positive if $m < b$. It is in this case, in fact, that a population explosion occurs in the absence of the saturation term.

In order to study the stability of this new solution, we consider a fairly small neighbourhood of x_0 such that, if

$$x = x_0 + \xi$$

we can write, neglecting higher-order terms in ξ

$$x^2 \approx x_0^2 + 2x_0 \xi$$

We then have

$$d\xi/dt = -\frac{bx_0}{A} \xi \qquad (5.12)$$

Proceeding in an analogous way in a fairly close neighbourhood around the origin, we have

$$d\xi/dt = (b - m) \, \xi \qquad (5.13)$$

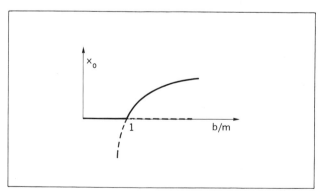

Fig. 5.3 - Continuous transition of "transcritical" type induced in Eq. (5.11) by changing the control parameter b/m. The stable solutions are represented by the solid lines, whereas the dashed lines indicate instability.

It is then clear that, when $b > m$, x_0 is a stable solution and the origin is an unstable solution. In the case $b < m$ both solutions invert their stability characteristics.

Let us now consider b/m as a "control" parameter in Eq. (5.11) while keeping the value of A fixed. When $b/m = 1$, there is thus a bifurcation; the inevitable presence of fluctuations gives rise, by increasing b/m, to a "continuous transition" of the type shown in Fig. 5.3.

This type of bifurcation differs from that discussed above, as in this case the equilibrium points are always two, whatever the value of the control parameter (even though one of them may lose physical meaning when it becomes negative). The two equilibrium points coincide when $b/m = 1$ and the stability characteristics are exchanged when the control parameter passes through the value 1. This type of bifurcation is called "transcritical".

We shall now bring a new hypothesis into our problem, by supposing that the mortality of the population considered essentially depends upon the activity of a certain "predatory" species, whose population, y, for simplicity may be assumed constant (i.e. stabilised by its interaction with other variables not considered explicitly in the present model).

The preceding analytical expression for the death term, mx, may then be interpreted as a description of the particular case in which the "preys" are quite scarce, so that the death rate is proportional to the number of encounters among the members of the two populations: $mx = Syx$. When, however, the value of x becomes very large, it is reasonable to suppose that the previous expression will be modified in the sense that the quantity of resources consumed by the population y will no longer depend upon x.

A simple death term which for small values of x is proportional to x and for large values of x is independent of x is of the type

$$\frac{Syx}{1 + x/B}$$

Eq. (5.11) can then be rewritten as

$$\frac{dx}{dt} = bx \left(1 - \frac{x}{A}\right) - \frac{Syx}{1 + x/B} \tag{5.14}$$

The "generalised force" on the r.h.s. may be considered as deriving from a potential (it must by kept in mind that $x > 0$)

$$V(x) = -\frac{b}{2} x^2 + \frac{b}{3A} x^3 + SyB \left[x - B \ln (x + B)\right] \tag{5.15}$$

Besides the point of equilibrium $x = 0$ now we also have two possible solutions given by the second degree equation

$$x_0^2 - (A - B) x_0 - AB \left(1 - \frac{Sy}{b}\right) = 0$$

from which

$$x_0^{(\pm)} = \frac{A - B}{2} \pm \frac{1}{2} \left[(A - B)^2 + 4AB \left(1 - \frac{Sy}{b}\right)\right]^{1/2} \tag{5.16}$$

Let us first of all consider the case $A/B > 1$. Physically, this involves the intervention of the self-limitation mechanism of species x only after the limitation mechanism due to the predators has fully played its part. For the discussion that follows, it will be useful to note here that the argument of the square root term is positive for

$$\frac{b}{Sy} > \frac{4AB}{(A+B)^2}$$

whereas when b/Sy equals this threshold value the two solutions $x_0^{(\pm)}$ coincide and are equal to $(A - B)/2$. The solution x_0^- remains positive (and therefore physically meaningful) as long as the addend $4AB(1 - Sy/b)$ is negative, and therefore when $b/Sy < 1$.

Considering only positive solutions, we can therefore divide the field of variability of the parameter b/Sy into three distinct regions.

1) $0 < b/Sy < 4AB/(A + B)^2$: the only solution of the equation considered is $x = 0$. This solution is stable and the potential (5.15) is in this case of the kind shown in Fig. 5.4a.
2) $4AB/(A + B)^2 < b/Sy < 1$: the equation allows three solutions, $0, x_0^-, x_0^+$. It is easily shown that 0 and x_0^+ are stable solutions, while x_0^- is unstable. The potential in this case takes the form shown in Fig. 5.4b.
3) $b/Sy > 1$: in this case there are only two solutions, $x = 0$ (which becomes unstable) and x_0^+ (which remains stable). The shape of the potential is given in Fig. 5.4c.

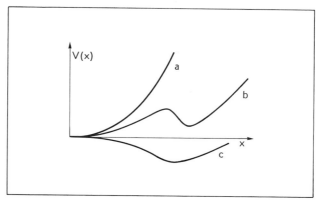

Fig. 5.4 - Different shapes of the potential (5.15) corresponding to various values of the control parameter b/Sy.

The bifurcation diagram for Eq. (5.14) (assuming b/Sy to be a control parameter) is as shown in Fig. 5.5, where the stable solutions are indicated with a continuous line and the unstable ones with a broken line.

We note here that on varying the parameter b/Sy, at the value $b/Sy = 4AB/(A + B)^2$ there is a bifurcation process of a different type than that seen previously. In particular, the stable solution for $b/Sy < 1$ remains stable even for $b/Sy > 1$. Let us assume first of all that the system represented by Eq. (5.14) is initially in the $x = 0$ configuration. In the presence of sufficiently small perturbations, the system remains in that configuration until the value of b/Sy reaches 1. At this point, since the equilibrium configuration becomes unstable, even a small perturbation is sufficient to allow the system to pass over to the configuration x_0^+.

If, however, b/Sy decreases, starting from an initial value greater than 1, the system remains in the configuration x_0^+ until it reaches the value

$$b/Sy = 4AB/(A + B)^2$$

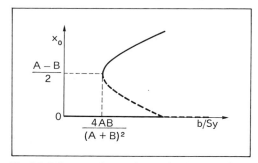

Fig. 5.5 - Bifurcation diagram for Eq. (5.14) taking b/Sy as a control parameter.

at which point it falls to $x = 0$.

We thus have a "hysteresis" phenomenon with "discontinuous transitions" where the discontinuity ("catastrophe") in the value of the equilibrium position occurs even for small variations in the control parameter.

Up to this point we have considered the case $A/B > 1$; if we now let A/B decrease to 1, we have

$$\frac{4AB}{(A+B)^2} \longrightarrow 1$$

If $A/B \ll 1$, we return to the starting equation with $Sy = m$. Taking A/B as a second control parameter, we can define an "equilibrium surface" as a function of the parameters b/Sy and A/B. This surface presents a "fold" whose extremes are projected upon the "control plane" in the cusp-shaped curve shown in Fig. 5.6.

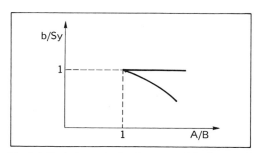

Fig. 5.6 - Projection onto the plane spanned by the control parameters of the equilibrium surface for Eq. (5.14).

A three-dimensional representation of the equilibrium surface is given in Fig. 5.7. Once again we can remark that in nonlinear systems the transitions (and thus the history of the systems themselves) critically depend upon the presence of fluctuations, as well as upon the structure of the equations used for their representation.

For simplicity's sake, we have limited our investigation to a first-order equation involving only two control parameters. In more general cases, there may be equilibrium hypersurfaces with complex folds which are more difficult to represent intuitively. Their study is the subject of "catastrophe theory", into which we will not enter here.

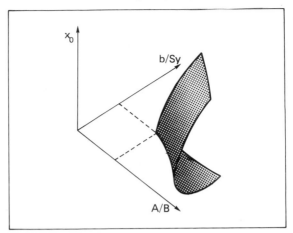

Fig. 5.7 - The equilibrium surface for Eq. (5.14).

It is worth noting that important examples of discontinuous transitions can be found in physics, such as those related to the Van der Waals equation or to the theory of ferromagnetism. Usually, however, their treatment is not developed starting from differential equations which the order parameters obey, but rather starting from the study of a state function ("equilibrium phase transitions").

5.3 EQUILIBRIUM AND STABILITY IN SECOND-ORDER SYSTEMS

We will now illustrate some of the fundamental concepts relating to differential systems of the following type:

$$\dot{\vec{x}} = \vec{F}(\vec{x}, \vec{P})$$

where $\vec{x} = (x_1, x_2, \ldots x_n)$ and $\vec{F} = (F_1, F_2, \ldots F_n)$ are n-dimensional vectors and $\vec{P} = (P_1, P_2, \ldots P_m)$ is an m-dimensional vector of parameters. Let us assume, for simplicity, that $\vec{F}$ does not explicitly depend upon time (autonomous systems); actually, this limitation is not of fundamental importance. It is always possible, in fact, to come back to this particular case by increasing the dimensions of the two vectors $\vec{x}$ and $\vec{F}$ by one and adding the equation

$$\dot{x}_{n+1} = 1$$

For simplicity, we will limit ourselves to two-dimensional systems, and to render the treatment more intuitive we will concentrate on the case where the different terms allow an immediate physical interpretation.

To this end, let us consider an extension of the previous problem, by taking into account the dynamics of the "predators" as well as that of the "preys". Along with this extension, however, we also include a certain simplification. That is, we will ignore (at least to begin with) the quadratic saturation term and assume that the mortality term for the "preys" is proportional to the product of the two populations. This latter hypothesis, as we recall, corresponds to a situation of relative scarcity of the "preys".

Representing the populations of "predators" and "preys" with x_1 and x_2 respectively, the system of equations (Lotka-Volterra equations) may be written as

$$\frac{dx_1}{dt} = -\alpha_1 x_1 + \lambda_1 x_1 x_2$$

$$\frac{dx_2}{dt} = \alpha_2 x_2 - \lambda_2 x_1 x_2$$

(5.17)

The physical meaning of the four coefficients on the r.h.s.'s of these equations is immediate.

The coefficient α_1 ($\alpha_1 > 0$) defines the mortality rate of species 1 (the "predators"); i.e., in the absence of "preys" they would undergo an exponential decay with a time constant $1/\alpha_1$.

The coefficient α_2 ($\alpha_2 > 0$) defines the birth rate of species 2 (the "preys"), and this, in its turn, depends upon the availability of adequate resources (the dynamics of which is not directly contemplated by the model). It is assumed that, in the absence of "predators", the "preys" would undergo an exponential explosion with a time constant $1/\alpha_2$.

The term $-\lambda_2 x_1 x_2$ represents the decrease of the "preys" due to the presence of the "predators", while the term $\lambda_1 x_1 x_2$ (which is of opposite sign to the previous term, as λ_1, $\lambda_2 > 0$) represents the increase of the "predators" due to the presence of the "preys".

As we can see, this is a particular case of a relationship between species 1 and 2. Different situations would be represented with different signs in the rectangular terms. For example, two positive signs would indicate a situation of "symbiosis", whereas two negative signs (possibly associated with the presence of other terms having a suitable physical meaning) would correspond to a situation of "competition", in which species 1 and 2 depend upon the same resources for their survival.

The equilibrium states for this system of two first-order differential equations are given by the algebraic system

$$x_{01}(-\alpha_1 + \lambda_1 x_{02}) = 0$$

$$x_{02}(\alpha_2 - \lambda_2 x_{01}) = 0$$

which allows the solutions $(0, 0)$ and $(\alpha_2/\lambda_2, \alpha_1/\lambda_1)$.

First of all, let us study the stability of the solution $(0, 0)$ with the linearisation method already seen above, which obviously must be extended to the two-dimensional case. If $x_1 = x_{01} + \xi_1$ and $x_2 = x_{02} + \xi_2$, developing the calculations and putting $x_{01} = x_{02} = 0$, we obtain

$$\frac{d\xi_1}{dt} = -\alpha_1 \xi_1$$

$$\frac{d\xi_2}{dt} = \alpha_2 \xi_2$$

(5.18)

The point of equilibrium $(0, 0)$ is thus unstable, because one of the two variables (namely, ξ_2) tends to move away indefinitely from the origin under the effect of arbitrarily small perturbations.

In the plane (ξ_1, ξ_2) the trajectories are of the type shown in Fig. 5.8a. Fig. 5.8b shows the potential $V(\xi_1, \xi_2)$ such that

$$\frac{d\xi_1}{dt} = -\frac{\partial V}{\partial \xi_1} \qquad \frac{d\xi_2}{dt} = -\frac{\partial V}{\partial \xi_2}$$

This potential evidently has the form

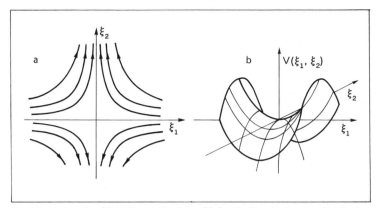

Fig. 5.8 - Trajectories (a) and potential shape (b) for the linearised system (5.18) when the unstable equilibrium state is a saddle point (α_1, $\alpha_2 > 0$).

$$V(\xi_1, \xi_2) = \frac{\alpha_1}{2}\xi_1^2 - \frac{\alpha_2}{2}\xi_2^2 \qquad (5.19)$$

Observing Fig. 5.8, one can clearly see why a point of unstable equilibrium of this type is called a "saddle point".

Let us now note that, if we allow the possibility of α_1 changing from positive to negative values, the origin, which initially is a saddle point, becomes a "node".

The term "node" is used for a point of equilibrium which corresponds to linearised equations of the type

$$\frac{d\xi_1}{dt} = a\,\xi_1$$

$$\frac{d\xi_2}{dt} = b\,\xi_2 \qquad (5.20)$$

with a and b of the same sign. In our case, we have a stable node if $\alpha_1 > 0$ and $\alpha_2 < 0$, and an unstable node if $\alpha_1 < 0$ and $\alpha_2 > 0$.

The potential V such that

$$\frac{d\xi_1}{dt} = -\frac{\partial V}{\partial \xi_1} \quad \text{and} \quad \frac{d\xi_2}{dt} = -\frac{\partial V}{\partial \xi_2}$$

in this case, as in the saddle point case, has the form

$$V = \frac{\alpha_1}{2}\xi_1^2 - \frac{\alpha_2}{2}\xi_2^2 \qquad (5.21)$$

Fig. 5.9 shows the trajectories in the plane (ξ_1, ξ_2) and the potential $V(\xi_1, \xi_2)$ for a stable node ($\alpha_1 > 0, \alpha_2 < 0$).

Let us now return to the case where α_1, $\alpha_2 > 0$ and study the stability of the point of equilibrium (α_2/λ_2, α_1/λ_1). Linearising the system (5.17), we obtain:

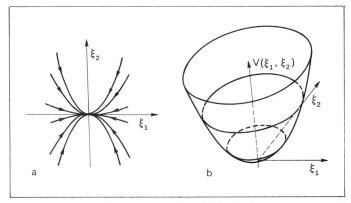

Fig. 5.9 - Trajectories (a) and potential shape (b) for the linearised system (5.18)
when the stable equilibrium point is a stable node ($\alpha_1 > 0$, $\alpha_2 < 0$).

$$\frac{d\xi_1}{dt} = \alpha_2 \frac{\lambda_1}{\lambda_2} \xi_2$$

$$\frac{d\xi_2}{dt} = -\alpha_1 \frac{\lambda_2}{\lambda_1} \xi_1$$

(5.22)

Differentiating the first equation with respect to time and substituting in it the expression for
$d\xi_2/dt$ obtained from the second equation, we have:

$$\frac{d^2\xi_1}{dt^2} + \alpha_1 \alpha_2 \xi_1 = 0$$

which is the equation for harmonic motion. The same is true for ξ_2.

Consequently, the representative point of the system (5.22) in the plane (ξ_1, ξ_2) performs
small harmonic oscillations about the point of equilibrium $(\alpha_2/\lambda_2, \alpha_1/\lambda_1)$. This point of equilibrium is therefore stable, because for $t \to \infty$ the point which represents the system always remains within a close neighbourhood. It is not, however, asymptotically stable, because it is not
possible to assert that the representative point of the system tends towards the point of equilibrium for $t \to \infty$.

The point of equilibrium in this case is a "center" and the trajectories in the plane (ξ_1, ξ_2)
are of the type shown in Fig. (5.10).

The "center" is a point of equilibrium with very different characteristics from those of
"saddles" and "nodes". In particular, it is not possible to define a potential V (which would
require the equality between the second mixed derivatives), but a constant of motion can be
easily individuated.

By starting from the linearised equations (5.22), multiplying the first by $2\xi_1 \alpha_1 (\lambda_2/\lambda_1)$ and
the second by $2\xi_2 \alpha_2 (\lambda_1/\lambda_2)$ and adding we obtain

$$\frac{d}{dt}\left[\alpha_1 \frac{\lambda_2}{\lambda_1} \xi_1^2 + \alpha_2 \frac{\lambda_1}{\lambda_2} \xi_2^2\right] = 0$$

(5.23)

Introducing suitable normalised variables:

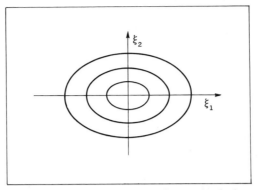

**Fig. 5.10 - Trajectories for the linearised system (5.22) whose
equilibrium point is a "center" ($\alpha_1, \alpha_2 > 0$).**

$$\xi_1' = \left(\alpha_1 \frac{\lambda_2}{\lambda_1} \right)^{1/2} \xi_1$$

$$\xi_2' = \left(\alpha_2 \frac{\lambda_1}{\lambda_2} \right)^{1/2} \xi_2$$

this equation can be rewritten

$$\xi_1'^2 + \xi_2'^2 = \text{const.}$$

and therefore the representative point of the system describes a circumference in the plane (ξ_1', ξ_2'), which corresponds to an ellipse in the plane (ξ_1, ξ_2).

We will now pass to polar coordinates, putting

$$\xi_1' = \rho \cos \vartheta \qquad \xi_2' = \rho \sin \vartheta$$

and recalling that

$$\xi_1' \dot{\xi}_2' - \xi_2' \dot{\xi}_1' = \rho^2 \dot{\vartheta}$$

$$\xi_1' \dot{\xi}_1' + \xi_2' \dot{\xi}_2' = \frac{1}{2} \frac{d}{dt} (\rho^2) = \rho \dot{\rho}$$

The system of equations (5.22) in polar coordinates then becomes

$$\frac{d\rho}{dt} = 0$$

$$\frac{d\vartheta}{dt} = -(\alpha_1 \alpha_2)^{1/2} = \text{const.}$$

(5.24)

As we can see, in polar coordinates it becomes possible to define a potential V which does not depend upon ρ. We thus have a situation in which ϑ varies with a constant angular velocity which depends only upon the parameters of the differential equations, while ρ is a constant whose value depends upon the initial conditions. The equilibrium value of ρ is therefore neutral. We note, however, that if we leave a strictly deterministic scheme, this characteristic gives rise to an extreme sensitivity to fluctuations.

Let us now consider a slightly different system, introducing into the second equation (5.17) a quadratic term of the type $-\gamma x_2^2$ ($\gamma > 0$). The physical meaning of this term, in the context of our example, is immediately clear: it expresses a "self-limitation" of the growth of the "preys" even in the absence of "predators". This "self-limitation", in its turn, may be related, for example, to a scarcity of the resources necessary for the survival of "preys" when they become very numerous.

It is immediately verified that the introduction of the quadratic term in the second equation (5.17) does not modify the stability characteristics of the origin, which remains a saddle point, and therefore unstable.

We will now study the new point of equilibrium different from the origin, the coordinates of which are given by the algebraic equations

$$-\alpha_1 + \lambda_1 x_{02} = 0$$

$$\alpha_2 - \lambda_2 x_{01} - \gamma x_{02} = 0$$

We then have

$$x_{01} = \frac{\alpha_2 - \gamma \alpha_1 / \lambda_1}{\lambda_2}$$

$$x_{02} = \alpha_1 / \lambda_1$$

The expression found for x_{01} is positive if $\lambda_1 \alpha_2 > \gamma \alpha_1$, i.e. if γ is fairly small. For reasons of space we will consider only this case here. The linearised equations about the point of equilibrium are

$$\frac{d\xi_1}{dt} = \frac{1}{\lambda_2} (\lambda_1 \alpha_2 - \gamma \alpha_1) \xi_2$$

$$\frac{d\xi_2}{dt} = -\alpha_1 \frac{\lambda_2}{\lambda_1} \xi_1 - \gamma \frac{\alpha_1}{\lambda_1} \xi_2$$

(5.25)

The effect of the quadratic term on the linearised equations (5.25) is twofold: in the first equation the coefficient of ξ_2 is modified (this coefficient, however, remains positive as long as the solution holds a physical significance); in the second equation the term $-\gamma(\alpha_1/\lambda_1)\xi_2$ appears. As before, we arrive at a second-order equation by differentiating the second equation (5.25) with respect to time and substituting the expression for $d\xi_1/dt$ which is obtained from the first equation. We obtain

$$\frac{d^2\xi_2}{dt^2} + \gamma \frac{\alpha_1}{\lambda_1} \frac{d\xi_2}{dt} + \frac{\alpha_1}{\lambda_1} (\lambda_1 \alpha_2 - \gamma \alpha_1) \xi_2 = 0$$

which is the equation for a damped harmonic motion. We can therefore state that the effect of introducing the quadratic term does not only modify the frequency, but also introduces a damping.

Solving the previous second-order equation, it can, in fact, be seen that the variable ξ_2 has a behaviour of the type

$$\xi_2 = A e^{-\gamma/2(\alpha_1/\lambda_1)t} \sin(\omega t + \varphi)$$

where

$$\omega^2 = \frac{\alpha_1}{\lambda_1} (\lambda_1 \alpha_2 - \gamma \alpha_1) - \frac{\gamma^2}{4} \frac{\alpha_1^2}{\lambda_1^2}$$

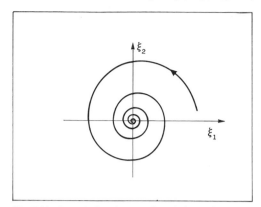

Fig. 5.11 - A typical trajectory described by the system (5.25), when $\gamma(\alpha_1/\lambda_1) > 0$, towards a stable focus.

An analogous equation is valid also for ξ_1.

The presence of the damping ensures that the point of equilibrium is not only stable but also asymptotically stable. In fact, for $t \to \infty$, both ξ_1 and ξ_2 tend towards zero.

The circumference (or, more generally, the ellipse) described previously by the representative point of the system (5.22) in the plane (ξ_1, ξ_2) now becomes, for the system (5.25), a spiral. This statement, however, may be easily demonstrated only after having effected a suitable linear transformation on the variables ξ_1 and ξ_2, as we shall see later on.

The point of equilibrium thus defined is called a "focus" (see Fig. 5.11). It is asymptotically stable if $\gamma(\alpha_1/\lambda_1) > 0$ and unstable if $\gamma(\alpha_1/\lambda_1) < 0$. In the light of the previous considerations, we can state that the center is a very particular case of focus in which the coefficient of damping (or, on the contrary, of moving away from equilibrium) $\gamma(\alpha_1/\lambda_1)$ is zero. Taking $\gamma(\alpha_1/\lambda_1)$ as a parameter, the center may be considered, for the linearised system (5.25), as a transition configuration between a stable and an unstable focus when the parameter passes through zero. Later we will see that, for nonlinear systems, more complex transitions are possible which give rise to undamped oscillating solutions leading to equilibrium trajectories in the phase plane ("limit cycles") which do not correspond to equilibrium points of the "center" type.

5.4 STABILITY OF EQUILIBRIUM POINTS AND CANONICAL TRANSFORMATIONS

We have seen that the study of the local stability of the equilibrium points for a system of two first-order differential equations can be carried out by linearising the equations around the different equilibrium points. So far, however, we have focused upon a particular example; it is necessary at this point to develop some more general considerations for the study of the local stability of second-order systems. A discussion of the overall behaviour of nonlinear systems will be deferred to Section 5.5.

Let us consider the system

$$\dot{x}_1 = f(x_1, x_2, \vec{P})$$

$$\dot{x}_2 = g(x_1, x_2, \vec{P}) \tag{5.26}$$

The vector of the parameters $\vec{P}$ is given particular emphasis because, as noted, in certain cases critical values of the parameters exist which can qualitatively change the characteristics and the stability of the equilibrium points. The equilibrium points of the system (5.26) are solutions of the system of nonlinear algebraic equations

$$f(x_1, x_2, \vec{P}) = 0$$
$$g(x_1, x_2, \vec{P}) = 0 \qquad (5.27)$$

In the phase plane (x_1, x_2), they correspond to the intersections of the two curves $f = 0$ and $g = 0$. Let $E_i = (x_{1i}(\vec{P}), x_{2i}(\vec{P}))$ be the equilibrium points found in this manner. Since the algebraic system (5.27) is non linear, it is impossible to determine the number of the equilibrium points without a knowledge of the system (5.27).

The linearisation procedure around each point involves the construction of the Jacobian matrix

$$\mathbf{J} = \begin{Vmatrix} \dfrac{\partial f}{\partial x_1} & \dfrac{\partial f}{\partial x_2} \\[2ex] \dfrac{\partial g}{\partial x_1} & \dfrac{\partial g}{\partial x_2} \end{Vmatrix} \qquad (5.28)$$

and the study of the system

$$\dot{\vec{\xi}} = \mathbf{J}\,\vec{\xi} \qquad (5.29)$$

where $\vec{\xi} = (\xi_1, \xi_2)$ and the matrix $\mathbf{J}$ is calculated at the equilibrium point considered. For the linear system (5.29), the only equilibrium point is the origin; moreover, the linearity allows one to solve the system and to study the dependence of the solutions and the stability of the equilibrium points upon the parameter vector $\vec{P}$.

With this aim in mind, by analogy with first-order differential equations, we can look for the conditions under which the system (5.29) allows solutions of the type

$$\vec{\xi} = \vec{u}_\lambda\, e^{\lambda t} \qquad (5.30)$$

On substitution, we are led to the system of algebraic equations

$$(\mathbf{J} - \lambda \mathbf{I})\,\vec{u}_\lambda = 0 \qquad (5.31)$$

where $\mathbf{I}$ is the identity matrix (or unitary diagonal matrix). This is a homogeneous system, which can have non-zero solutions only if the determinant of the coefficients is zero. In other words, the values of λ for which the system (5.29) allows non-zero solutions of the type (5.30) are the eigenvalues of the Jacobian matrix (5.28).

It is then necessary to calculate the roots of the characteristic polynomial

$$\lambda^2 - \lambda \left(\frac{\partial f}{\partial x_1} + \frac{\partial g}{\partial x_2} \right) + \left(\frac{\partial f}{\partial x_1} \frac{\partial g}{\partial x_2} - \frac{\partial f}{\partial x_2} \frac{\partial g}{\partial x_1} \right) = 0 \qquad (5.32)$$

where, as already mentioned, the partial derivatives are calculated at the particular equilibrium point of the system (5.26) for which the local stability is being studied.

For simplicity's sake, let us assume that

$$\frac{\partial f}{\partial x_1} + \frac{\partial g}{\partial x_2} = S \quad \text{and} \quad -\frac{\partial f}{\partial x_2} \frac{\partial g}{\partial x_1} + \frac{\partial f}{\partial x_1} \frac{\partial g}{\partial x_2} = Q \qquad (5.33)$$

so that Eq. (5.32) can be written

$$\lambda^2 - S\lambda + Q = 0 \tag{5.34}$$

From Eq.(5.34) two eigenvalues can be obtained:

$$\lambda_1 = \frac{S + \sqrt{S^2 - 4Q}}{2}$$

$$\lambda_2 = \frac{S - \sqrt{S^2 - 4Q}}{2} \tag{5.35}$$

The discriminant $(S^2 - 4Q)$ will be indicated by Δ :

$$\Delta = S^2 - 4Q = \left(\frac{\partial f}{\partial x_1} - \frac{\partial g}{\partial x_2} \right)^2 + 4 \frac{\partial f}{\partial x_2} \frac{\partial g}{\partial x_1} \tag{5.36}$$

If S and Q are taken as parameters, the discussion of the eigenvalues is particularly simple and can be summarised graphically as in Fig. 5.12.

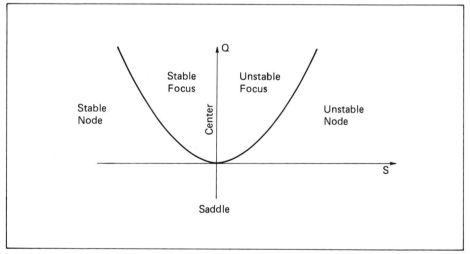

Fig. 5.12 - **Graphical representation of the dependence of the eigenvalues of Eq. (5.31) upon the values of S and Q.**

The parabola $Q = S^2/4$ corresponds to the surface which, in parameter space, separates the region of real eigenvalues from that of complex eigenvalues. At $Q = S^2/4$ the eigenvalues are real and coincident

$$\lambda_1 = \lambda_2 = \frac{1}{2} \left(\frac{\partial f}{\partial x_1} + \frac{\partial g}{\partial x_2} \right)$$

Regarding the elements of the Jacobian matrix, it can be seen that in the region of complex eigenvalues we must have

$$\left(\frac{\partial f}{\partial x_1} - \frac{\partial g}{\partial x_2} \right)^2 < -4 \frac{\partial f}{\partial x_2} \frac{\partial g}{\partial x_1}$$

so that $\partial f/\partial x_2$ and $\partial g/\partial x_1$ must necessarily be of opposite sign. The half-axis $S = 0, Q > 0$ corresponds, in the parameter space, to the locus of the purely imaginary eigenvalues for the linear system (5.29).

We must now find a connection between the nature of the eigenvalues and the dynamic behaviour of the solutions of the system (5.29) which, due to the linearity of the system itself, may be written

$$\vec{\xi} = c_1 \vec{u}_1 e^{\lambda_1 t} + c_2 \vec{u}_2 e^{\lambda_2 t}$$

In order to study this problem, it is worth carrying out a linear transformation on the vector $\vec{\xi}$

$$\vec{\xi} = S \vec{\xi}' \tag{5.37}$$

so that the nature of the singular points is not changed The system (5.29) is then transformed into the equivalent

$$\dot{\vec{\xi}}' = S^{-1} J S \vec{\xi}' = J' \vec{\xi}' \tag{5.38}$$

The transformation matrix S is thus to be chosen so that J' has a particularly simple form. The most suitable "canonical transformation" will depend upon the nature of the singular points. Let us now consider the case of real eigenvalues (which, for simplicity, are assumed to be non-coincident) separately from the case of complex conjugate eigenvalues.

In the former case ($\Delta > 0$), solving Eq. (5.31), the eigenvector pairs (defined apart from a multiplicative constant) corresponding to the eigenvalues λ_1 and λ_2 are:

$$\vec{u}_1 = \begin{pmatrix} \dfrac{\partial f}{\partial x_2} \\ \lambda_1 - \dfrac{\partial f}{\partial x_1} \end{pmatrix} \qquad \vec{u}_2 = \begin{pmatrix} \dfrac{\partial f}{\partial x_2} \\ \lambda_2 - \dfrac{\partial f}{\partial x_1} \end{pmatrix}$$

for $\partial f/\partial x_2 \neq 0$, or $\tag{5.39}$

$$\vec{u}_1 = \begin{pmatrix} \lambda_1 - \dfrac{\partial g}{\partial x_2} \\ \dfrac{\partial g}{\partial x_1} \end{pmatrix} \qquad \vec{u}_2 = \begin{pmatrix} \lambda_2 - \dfrac{\partial g}{\partial x_2} \\ \dfrac{\partial g}{\partial x_1} \end{pmatrix}$$

for $\partial g/\partial x_1 \neq 0$.

In the particular case $\partial f/\partial x_2 = \partial g/\partial x_1 = 0$, the Jacobian matrix is already diagonal, and the eigenvectors are

$$\begin{pmatrix} 1 \\ 0 \end{pmatrix} \quad \text{and} \quad \begin{pmatrix} 0 \\ 1 \end{pmatrix}$$

When the matrix

$$S = \left\| \begin{array}{cc} \dfrac{\partial f}{\partial x_2} & \dfrac{\partial f}{\partial x_2} \\[2mm] \lambda_1 - \dfrac{\partial f}{\partial x_1} & \lambda_2 - \dfrac{\partial f}{\partial x_1} \end{array} \right\| \tag{5.40}$$

made up with the eigenvectors in the form (5.39) is used for the canonical transformation, one obtains

$$S^{-1} J S = \left\| \begin{array}{cc} \lambda_1 & 0 \\ 0 & \lambda_2 \end{array} \right\| \tag{5.41}$$

The system (5.29) then becomes

$$\begin{aligned} \dot{\xi}'_1 &= \lambda_1 \xi'_1 \\ \dot{\xi}'_2 &= \lambda_2 \xi'_2 \end{aligned} \tag{5.42}$$

The eigenvalues $\lambda_{1,2}$ take the form (5.35), and the discussion of their sign can proceed by analogy with that developed in Section 5.3. In particular, if $Q < 0$ the two eigenvalues have opposite sign and a saddle point is obtained. If $Q > 0$ we have a stable node or an unstable node according to whether $S < 0$ or $S > 0$, respectively.

When $\Delta < 0$, the eigenvalues are complex conjugates

$$\begin{aligned} \lambda_1 &= c + i\,\omega \\ \lambda_2 &= c - i\,\omega \end{aligned} \tag{5.43}$$

Correspondingly, also the eigenvectors are complex conjugates

$$\begin{aligned} \vec{u}_1 &= \vec{a} + i\vec{b} \\ \vec{u}_2 &= \vec{a} - i\vec{b} \end{aligned} \tag{5.44}$$

where $\vec{a}$ and $\vec{b}$ are solutions of the linear system

$$\begin{aligned} J\vec{a} &= c\,\vec{a} - \omega\vec{b} \\ J\vec{b} &= \omega\vec{a} + c\,\vec{b} \end{aligned}$$

In the case of a two-dimensional system, in which we are interested here, on substituting Eq. (5.43) into the explicit expressions (5.39) we obtain:

$$\vec{u}_{1,2} = \begin{bmatrix} \dfrac{\partial f}{\partial x_2} \\[3mm] \lambda_{1,2} - \dfrac{\partial f}{\partial x_1} \end{bmatrix} = \begin{bmatrix} \dfrac{\partial f}{\partial x_2} \\[3mm] c - \dfrac{\partial f}{\partial x_1} \end{bmatrix} \pm i \begin{bmatrix} 0 \\[3mm] \omega \end{bmatrix}$$

or

$$\vec{u}_{1,2} = \begin{bmatrix} \lambda_{1,2} - \dfrac{\partial g}{\partial x_2} \\[3mm] \dfrac{\partial g}{\partial x_1} \end{bmatrix} = \begin{bmatrix} c - \dfrac{\partial g}{\partial x_2} \\[3mm] \dfrac{\partial g}{\partial x_1} \end{bmatrix} \pm i \begin{bmatrix} \omega \\[3mm] 0 \end{bmatrix}$$

Using the matrix

$$\mathbf{S} = \| \vec{a}\ \vec{b} \|$$ (5.45)

for the canonical transformation (5.37), we obtain

$$\mathbf{S}^{-1}\mathbf{J}\mathbf{S} = \begin{Vmatrix} c & \omega \\ -\omega & c \end{Vmatrix}$$ (5.46)

In this case, the system (5.29) becomes

$$\begin{aligned} \dot{\xi}_1' &= c\,\xi_1' + \omega\,\xi_2' \\ \dot{\xi}_2' &= -\omega\,\xi_1' + c\,\xi_2' \end{aligned}$$ (5.47)

Changing to polar coordinates putting

$$\begin{aligned} \xi_1' &= \rho \cos \vartheta \\ \xi_2' &= \rho \sin \vartheta \end{aligned}$$

the system (5.47) becomes

$$\begin{aligned} \dot{\rho} &= c\,\rho \\ \dot{\vartheta} &= -\omega \end{aligned}$$ (5.48)

The stability analysis of this system is then immediate. As it can be seen, in fact, the equilibrium point (the origin) is asymptotically stable if $c < 0$ (stable focus), and unstable if $c > 0$ (unstable focus). If $c = 0$ the origin is a center.

5.5 STRUCTURE OF THE PHASE PLANE

A knowledge of the equilibrium points of the system (5.26) and the study of their local stability are not sufficient for the determination of the structure of the phase plane of the system itself. At this point of the analysis, in fact, we can simply state that the trajectories can intersect only at equilibrium points, and that the equilibrium points themselves are locally saddles, nodes, foci or centers.

In equilibrium points of node or focus type, an infinite number of trajectories can intersect, corresponding to different initial conditions (see Figs. 5.9 and 5.11), whereas only two trajectories intersect at saddle points: one is the "stable manifold", identified by the eigenvector of the linearised system (5.29) which corresponds to the negative eigenvalue, and the other is the "unstable manifold", identified by the eigenvector of the linearised system (5.29) which corresponds to the positive eigenvalue. These eigenvectors, however, are deformed on moving away from the region in which the linear approximation is valid (see Fig. 5.13a).

An overall image of the phase plane requires the determination of the "separatrices", which coincide with the eigenvectors in a sufficiently small region around the equilibrium points. The separatrices may close on themselves, as in the case shown in Fig. 5.13b ("homoclinic trajectories") or link different equilibrium points ("heteroclinic trajectories") as in Fig. 5.13c. They separate the "basins of attraction" of the different equilibrium points.

In certain cases, the solutions which correspond to initial conditions within a certain region may be attracted by the separatrices themselves (see Fig. 5.14).

As Fig. 5.15 shows, attractors of another type may exist, consisting of closed orbits which surround a focus and which may be travelled in a finite time (in contrast with the homoclinic orbits), thus becoming periodic orbits. Such attractors are called "limit cycles". In the particular

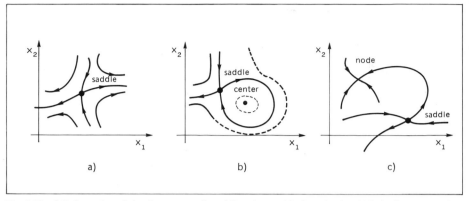

**Fig. 5.13 - a) Deformation of the eigenvectors of a saddle point outside the region in which the linear approxima-
tion is valid; b) an example of a homoclinic trajectory; c) an example of a heteroclinic trajectory.**

case shown in Fig. 5.15, the limit cycle is asymptotically stable. Unstable limit cycles, however,
may also occur.

In order to verify, in a simple way, that limit cycles can in fact occur, let us consider the
nonlinear system

$$\begin{aligned}
\dot{x}_1 &= -x_2 - x_1(x_1^2 + x_2^2 - 1) \\
\dot{x}_2 &= x_1 - x_2(x_1^2 + x_2^2 - 1)
\end{aligned} \tag{5.49}$$

Changing to polar coordinates by means of the transformation

$$x_1 = \rho \cos \vartheta \qquad\qquad x_2 = \rho \sin \vartheta$$

the system (5.49) becomes

$$\frac{d\rho}{dt} = \rho(1 - \rho^2)$$

$$\frac{d\vartheta}{dt} = 1 \tag{5.50}$$

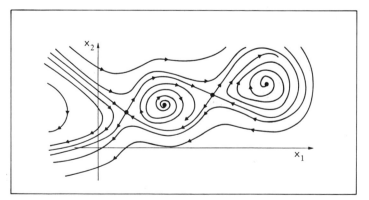

Fig. 5.14 - A possible structure of the phase plane for a second-order nonlinear system.

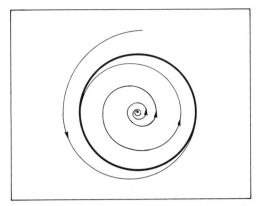

Fig. 5.15 - Example of an asymptotically stable limit cycle.

We have now two uncoupled equations, the first of which, as is already known, allows a stable equilibrium value $\rho_0 = 1$. This corresponds to a cyclic periodic motion around the origin for the system (5.49). We are, however, dealing with a type of motion which is completely different from that which corresponds to a center for linear systems. The amplitude of the limit cycle does not, in fact, depend upon the initial conditions, but only upon the equation parameters. The limit cycle acts as an "attractor" for trajectories which correspond to initial conditions which are not on the limit cycle itself. The determination of the limit cycle for the system (5.49) does not require a linearisation procedure, so that the result obtained has global and not just local validity.

Procedures of a general nature in the search for limit cycles are, however, unknown. The most powerful structured approach would seem to be that of "embedding" the system of interest, corresponding to defined values of the parameters, into a more general class of systems which depend upon a vector of parameters $\vec{P}$. As we shall see, the study of bifurcations as functions of the parameters allows, under quite general conditions, to predict the existence of a limit cycle, at least sufficiently close to the bifurcation itself.

5.6 BIFURCATION ANALYSIS

As mentioned above, the general form of the system (5.26) contemplates its dependence upon a vector of parameters. Following variations in these parameters, changes will occur not only in the coordinates of the equilibrium points of the system, but also in their number and in their stability characteristics. On examining Fig. 5.12, it is clear that significant bifurcations will occur when, following variations in the parameters, the representative point of the linearised system (5.29) in the plane (S, Q) crosses the line $S < 0$, $Q = 0$ or the line $S = 0$, $Q > 0$. In the first case, it is a real eigenvalue corresponding to the linearised system (5.29) which changes sign passing through zero; in the second case, the real part of a pair of complex conjugate eigenvalues changes sign.

Let us consider the first case. For simplicity, let us assume that the behaviour of the system (5.26) close to a bifurcation point may be expressed by means of a one-dimensional parameter μ, a function of the parameter vector $\vec{P}$. Let us then assume that for $\mu = \mu_0$, corresponding to an equilibrium point $(x_{01}(\mu_0), x_{02}(\mu_0))$, there is a real zero-valued eigenvalue, and in particular that $\lambda_1 = 0$. Let us change the coordinates of the nonlinear system (5.26):

$$x_1' = x_1 - x_{01}(\mu)$$

$$x_2' = x_2 - x_{02}(\mu)$$

and apply the canonical transformation (5.40) to the resulting system in x_1', x_2'. As we consider the nonlinear system, besides the canonical terms of (5.42) we shall obtain, in each of the resulting equations, additional terms. That is, we shall have a system of the following type:

$$\dot{\xi}_1 = \lambda_1(\mu)\,\xi_1 + f_1(\xi_1, \xi_2, \mu)$$

$$\dot{\xi}_2 = \lambda_2(\mu)\,\xi_2 + f_2(\xi_1, \xi_2, \mu)$$

(5.51)

Let us now assume, for example, that $\lambda_2(\mu) < 0$. Since we know that $\lambda_1(\mu) \xrightarrow[\mu \to \mu_0]{} 0$, the variable ξ_2 may be considered a "slave" of ξ_1 for values of μ in a fairly small region around μ_0. If, however, $\lambda_2 > 0$, the same considerations are valid if the direction of time is inverted.

Applying the direct adiabatic elimination procedure, which consists in setting equal to zero the time derivative of the fast variable, or a more refined elimination procedure (see Chapter 3), it is then possible to study a single differential equation, which is usually nonlinear

$$\dot{\xi}_1 = f(\xi_1, \mu)$$

(5.52)

The "central manifold" in which the bifurcation is to be studied thus has only one dimension. In the particular case where Eq. (5.52) is of the type

$$\dot{\xi}_1 = K(\mu)\,\xi_1$$

the only effect produced by letting the parameter μ pass through μ_0 is a change in the stability of the equilibrium point. Usually, however, Eq. (5.52) also contains higher-order terms, so that different situations may occur. If, in particular, a quadratic term is present besides the linear one, as in the case of Eq. (5.11), the bifurcation is of the "transcritical" type, with an exchange of stability between two equilibrium points. When a third-order term is present, as in Eq. (5.9), we have a "pitchfork" bifurcation which (assuming $\lambda_2 < 0$) from one stable equilibrium point leads to three, of which two are stable and one unstable.

An adiabatic elimination procedure may also introduce an additive constant, $\alpha(\mu)$, into the equation for ξ_1. Let us assume, without any loss of generality, that $\alpha(\mu_0) = 0$ and $(\partial\alpha/\partial\mu)_{\mu=\mu_0} < 0$. Besides the additive constant and the linear term, usually Eq. (5.52) will contain, in this case, at least one second-order term. Since a suitable transformation of coordinates can eliminate the linear term, the simplest expression for Eq. (5.52) will then become

$$\dot{\xi}_1 = \alpha(\mu) - \xi_1^2$$

(5.53)

When α passes through zero, following the variation in μ, the real equilibrium points, which exist for $\mu < \mu_0$, disappear for $\mu > \mu_0$. Of these equilibrium points on the central manifold, one is unstable ($\xi_1 = -\sqrt{\alpha(\mu)}$) and the other is stable ($\xi_1 = \sqrt{\alpha(\mu)}$). Since we have assumed that $\lambda_2 < 0$, the latter correspond, in the phase plane, to a saddle and a stable node, respectively. For this reason, a bifurcation, which on the center manifold is described by an equation of the type (5.53), is called a "saddle-node" bifurcation.

Let us now consider the bifurcation where the real part of a pair of complex conjugate eigenvalues of the Jacobian matrix (5.28) vanishes. We will assume, also in this case, that the bifurcation takes place at $\mu = \mu_0$. The eigenvalues will then be of the type

$$\lambda_\pm = c(\mu) \pm i\,\omega(\mu)$$

with $c(\mu_0) = 0$. As in the previous case, the origin of the coordinates is translated into the equilibrium point $(x_{01}(\mu), x_{02}(\mu))$. The canonical transformation (5.45) is then applied to the resulting system, obtaining

$$\dot{\xi}_1 = c(\mu)\,\xi_1 + \omega(\mu)\,\xi_2 + f_1(\xi_1, \xi_2, \mu)$$
$$\dot{\xi}_2 = -\omega(\mu)\,\xi_1 + c(\mu)\,\xi_2 + f_2(\xi_1, \xi_2, \mu)$$

(5.54)

In this case, the central manifold in which the bifurcation is studied must necessarily be two-dimensional, because the conditions required for the adiabatic elimination of one of the two variables do not occur. Only in the case where

$$f_1(\xi_1, \xi_2, \mu) = f_2(\xi_1, \xi_2, \mu) = 0$$

(that is, when the system (5.26) itself is linear), the bifurcation simply corresponds to passing from a stable to an unstable focus (or vice versa) through a center. If, on the contrary, the system (5.54) is nonlinear, a more detailed analysis is required. In particular, the presence of non-zero terms, for $\mu = \mu_0$, in addition to the canonical terms, causes the equilibrium point for $\mu = \mu_0$ to be possibly different from a center.

A detailed analysis of this latter case is the object of the Hopf theorem, which provides the conditions for the birth or the disappearance of a limit cycle at the bifurcation.

Let us assume then that for $\mu = \mu_0$ there is a pair of purely imaginary eigenvalues for the linearised system, with $\omega(\mu_0) \neq 0$, $c(\mu_0) = 0$ and $(dc/d\mu)_{\mu=\mu_0} \neq 0$. When $(dc/d\mu)_{\mu=\mu_0} > 0$ (< 0) a transition occurs from a stable focus to an unstable one with increasing (decreasing) μ. Given these hypotheses, the Hopf theorem states that there are three (and only three) possible types of transitions from an unstable focus to a stable focus, according to the characteristics of the equilibrium point at $\mu = \mu_0$. If for $\mu = \mu_0$ the equilibrium point is a stable focus (on the basis, obviously, of an analysis which also takes into account f_1 and f_2 on the r.h.s. of Eq. (5.54)), then for $\mu > \mu_0$ the focus becomes unstable and, at the same time, a stable limit cycle arises whose amplitude, which is infinitesimal for an infinitesimal μ, grows as $(\mu - \mu_0)^{1/2}$ ("supercritical bifurcation"). If for $\mu = \mu_0$ the equilibrium point is an unstable focus, then for $\mu < \mu_0$ a stable focus occurs, surrounded by an unstable limit cycle, while for $\mu > \mu_0$ the limit cycle disappears ("subcritical bifurcation"). Finally, if for $\mu = \mu_0$ the equilibrium point is a center, there is a simple step from a stable to an unstable focus, without the presence of limit cycles, at least for fairly small values of $\mu - \mu_0$ ("critical bifurcation"). For $\mu = \mu_0$, however, an infinite number of periodic orbits occur in phase space which depend upon the initial conditions. Fig. 5.16 shows the three possible "Hopf bifurcations".

It is worth stressing once again that the results of the Hopf theorem are valid locally, in the sense that they refer to fairly small values of $\mu - \mu_0$. With increasing values of $\mu - \mu_0$, it may occur, e.g., that the amplitude of a stable limit cycle corresponding to a supercritical bifurcation increases indefinitely, or decreases to a point, or tends towards a homoclinic orbit.

As mentioned, a study of the equilibrium point for $\mu = \mu_0$ requires that the non-canonical terms of the system (Eq. 5.54) are also taken into account.

On this point, it is to be noted that the system in canonical form (5.54) may be reduced to an even easier "normal form" by a suitable nonlinear transformation of the variables

$$\tilde{\xi}_1 = \tilde{\xi}_1(\xi_1, \xi_2)$$
$$\tilde{\xi}_2 = \tilde{\xi}_2(\xi_1, \xi_2)$$

The system thus obtained can then be conveniently written in polar coodinates using the transformation

$$\tilde{\xi}_1 = \rho \cos \vartheta$$
$$\tilde{\xi}_2 = \rho \sin \vartheta$$

giving a result of the type

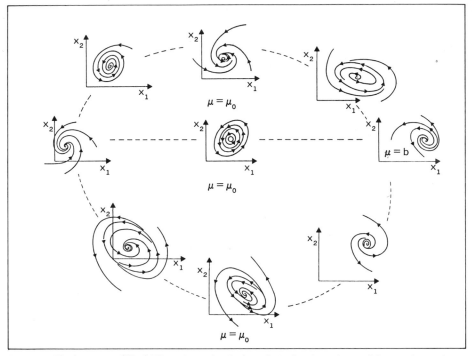

Fig. 5.16 - The three types of Hopf bifurcation under the hypothesis that the real part of the complex conjugate eigenvalues increases with the parameter μ.

$$\dot{\rho} = A\rho^3 + \text{higher-order terms}$$

$$\dot{\vartheta} = -\omega + B\rho^2 + \text{higher-order terms}$$

It can be immediately seen that the equilibrium point for $\mu = \mu_0$ is a stable focus if $A < 0$ and an unstable one if $A > 0$. For reasons of brevity, we have omitted the intermediate steps. It is, however, possible to write the coefficient A (which, from what has been said above, contains the essential information) in terms only of the functions which appear in Eq. (5.54), thus obtaining:

$$
\begin{aligned}
A = \frac{1}{16} & \left[\frac{\partial^3 f_1}{\partial \xi_1^3} + \frac{\partial^3 f_1}{\partial \xi_1 \partial \xi_2^2} + \frac{\partial^3 f_2}{\partial \xi_1^2 \partial \xi_2} + \frac{\partial^3 f_2}{\partial \xi_2^3} \right] - \\
& - \frac{1}{16\,|\omega(\mu_0)|} \left[\frac{\partial^2 f_1}{\partial \xi_1 \partial \xi_2} \left(\frac{\partial^2 f_1}{\partial \xi_1^2} + \frac{\partial^2 f_1}{\partial \xi_2^2} \right) - \right. \\
& - \frac{\partial^2 f_2}{\partial \xi_1 \partial \xi_2} \left(\frac{\partial^2 f_2}{\partial \xi_1^2} + \frac{\partial^2 f_2}{\partial \xi_2^2} \right) - \frac{\partial^2 f_1}{\partial \xi_1^2} \frac{\partial^2 f_2}{\partial \xi_1^2} + \\
& \left. + \frac{\partial^2 f_1}{\partial \xi_2^2} \frac{\partial^2 f_2}{\partial \xi_2^2} \right]
\end{aligned}
$$

(5.55)

where all the derivatives are to be evaluated at $\mu = \mu_0$. As mentioned previously, the equilibrium point is a stable focus (and we thus have a subcritical bifurcation) if $A > 0$, and an unstable focus (and we thus have a supercritical bifurcation) if $A > 0$. The $A = 0$ case requires the study of higher-order derivatives.

In order to illustrate an example of a Hopf bifurcation, let us now consider a prey-predator system slightly different from that represented by Eq. (5.17) regarding the dynamics of the prey. As mentioned above, the description of the prey-predator dynamics using the system (5.17) is only one of many possibilities: on changing the starting hypotheses, there are also changes in the corresponding equations.

Let us assume, in particular, that the growth of the prey population depends upon the frequency of encounters between elements of the same species (this could represent, for example, a reasonable model for sexual reproduction, given a certain equilibrium between the number of males and females). It is reasonable, in this case, that the growth term is a second-order rather than a first-order one. From this term, which grows faster than a first-order one for fairly numerous populations, we will subtract a higher-order term (third-order, for example), in order to avoid the non-physical effect of an unlimited growth. If the equation for the predators is the same as that considered previously, one then has a system of two differential equations which depend upon five parameters. In order to simplify the analysis of the Hopf bifurcation, it is convenient, as noted above, to reduce, as far as possible, the number of parameters. Towards this end let us assume that the populations and the time are suitably normalised, thus reducing the number of parameters to only two. In particular, it is assumed that the normalised equations are of the type:

$$\dot{x}_1 = x_1^2 (1 - x_1) - x_1 x_2$$
$$\dot{x}_2 = k (x_1 - \mu) x_2 \tag{5.56}$$

with $k, \mu > 0$. The equilibrium points for the system (5.56) are $(0, 0)$ $(1, 0)$, $(\mu, \mu (1 - \mu))$. Since, for physical reasons, it is assumed that $x_1, x_2 > 0$, the parameter μ can only vary between 0 and 1.

Instead of carrying out a complete analysis of the system (5.56), for the sake of brevity we will limit our considerations to the equilibrium point $(\mu, \mu (1 - \mu))$. The Jacobian matrix (5.28) for the system (5.56) is written as

$$J = \begin{Vmatrix} 2x_1 - 3x_1^2 - x_2 & -x_1 \\ kx_2 & kx_1 - k\mu \end{Vmatrix} \tag{5.57}$$

At the equilibrium point here considered, this matrix reduces to

$$J = \begin{Vmatrix} \mu(1 - 2\mu) & -\mu \\ k\mu(1 - \mu) & 0 \end{Vmatrix} \tag{5.58}$$

so that the eigenvalues (5.35) are

$$\lambda_{\pm} = \frac{1}{2}\{\mu(1 - 2\mu) \pm [\mu^2(1 - 2\mu)^2 - 4k\mu^2(1 - \mu)]^{1/2}\} \tag{5.59}$$

When $\mu = 1/2$ the eigenvalues (5.59) are purely imaginary: when μ passes through the value $1/2$ on going from higher to lower values, the real part of the eigenvalues passes from negative (stable focus) to positive (unstable focus) values. Thus the conditions of the Hopf theorem are satisfied, with $(dc/d\mu)_{\mu=\mu_0} < 0$, and the bifurcation may be studied as a function of the single parameter μ.

With the aim of applying the Hopf theorem, the first step is the transformation which brings the origin to the equilibrium point $(\mu, \mu(1 - \mu))$.

$$x_1' = x_1 - \mu \qquad x_2' = x_2 - \mu(1 - \mu)$$

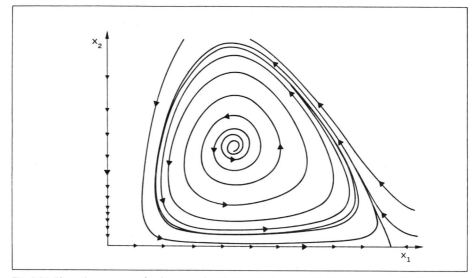

Fig. 5.17 -Phase plane structure for the system (5.56) determined numerically for $k = 1$ and $\mu = 0.465$.

In this way the following system is obtained:

$$\dot{x}'_1 = \mu(1 - 2\mu) x'_1 - \mu x'_2 - x'_1 x'_2 + x'^2_1 (1 - 3\mu) - x'^3_1$$
$$\dot{x}'_2 = k\mu(1 - \mu)x'_1 + kx'_1 x'_2 \tag{5.60}$$

The canonical transformation (5.45) must now be applied to the system (5.60). The result is given here for $\mu = 1/2$:

$$\dot{\xi}_1 = \sqrt{\frac{k}{8}}\, \xi_2 + f(\xi_1, \xi_2)$$

$$\dot{\xi}_2 = -\sqrt{\frac{k}{8}}\, \xi_1 + g(\xi_1, \xi_2) \tag{5.61}$$

where

$$f(\xi_1, \xi_2) = k\sqrt{\frac{k}{8}}\, \xi_1 \xi_2$$

$$g(\xi_1, \xi_2) = -\frac{k}{4} \xi_1 \xi_2 - \frac{1}{2}\sqrt{\frac{k}{8}}\, \xi_1^2 - \frac{k}{8} \xi_2^3 \tag{5.62}$$

Proceeding with the calculations, it is found that the coefficient A (Eq. (5.55)) has the value

$$A \big|_{\mu=1/2} = -\frac{k}{32} < 0 \tag{5.63}$$

For $\mu = 1/2$ there is thus a supercritical bifurcation, and for $\mu < 1/2$ (which corresponds to an unstable focus) the Hopf theorem predicts the existence of a stable limit cycle. Fig. 5.17

shows the structure of the phase plane (which can only be determined numerically) for the system (5.56) with $k = 1$ and $\mu = 0.465$. As can be seen, for these values a limit cycle does indeed exist. The Hopf theorem, as we know, does not provide the interval of values for which the limit cycle exists: this interval can only be determined numerically. Figure 5.17 also shows the behaviour of the system (5.56) in the neighbourhood of the other two equilibrium points, which are both unstable. It can be seen that the axes x_1 and x_2 both correspond to system trajectories.

5.7 HIGHER-ORDER SYSTEMS AND POINCARE' MAPS

As we have seen in second-order dynamical systems (planar systems) the possible "attractors" are fixed points, orbits connecting fixed points (heteroclinic or homoclinic orbits) and planar limit cycles. Such types of attractors are also present in higher-order systems. Besides these, other types of attractors are possible, due to the greater number of dimensions of the phase space. In particular, it is clear that possible limit cycles will no longer necessarily be planar, but may be curves in space, as shown in Fig. 5.18.

The limit cycle shown in Fig. 5.18, which may be considered as a closed curve on a toroidal surface, points to the possibility that the whole toroidal surface itself could constitute an attractor for a differential system of third (or higher)-order. Generally, then, it is reasonable that the attractors for a dynamical n-th-order system may take any integral number of dimensions between 0 (fixed points) and $n - 1$. We shall see below, however, that attractors of different types can also exist.

The bifurcation analysis carried out in Section 5.6 for planar systems can also be extended to higher-order systems. By means of a suitable transformation of variables, it is possible also in this case to develop the bifurcation analysis on a central manifold identified by the variables corresponding to eigenvalues whose real part vanishes at the bifurcation itself. The Hopf theorem may thus be formulated, on the central manifold, in the same way as for planar systems.

It can be seen at this point that in the systems of differential equations considered so far we have supposed that every trajectory is perfectly predictable, as long as the initial conditions are known with sufficient precision. We can consider this hypothesis as a translation into physical terms of the fundamental Cauchy theorem on the existence and uniqueness of the solutions. The validity of this theorem, however, does not imply its effective practical applicability, because it is not possible to define with infinite precision the initial conditions.

Besides, we know that it is, in general, impossible to analytically solve systems of even relatively simple nonlinear differential equations, so that it becomes necessary to integrate them

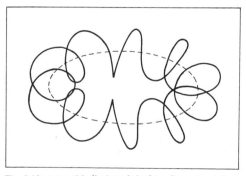

Fig. 5.18 - A possible limit cycle in three-dimensional phase
space.

numerically or analogically. This inevitably introduces a source of noise (truncation "noise", electronic noise) the effect of which essentially depends upon the sensitivity of the system of differential equations to the initial conditions. This sensitivity, in turn, may be better understood if it is referred to the structure of the phase space.

To clarify this concept, let us refer, for example, to the case of a simple pendulum. As is well-known, the latter may oscillate or rotate about its own axis, according to the initial conditions. In particular, the rotation of the pendulum clearly requires more energy than the oscillation. In the phase plane $(\vartheta, \dot{\vartheta})$, where ϑ is the angle of deflection, these two types of motion correspond to disjoint regions. If, therefore, we know that the system finds itself in a particular region, we can be sure of the type of motion that the pendulum will describe. Increasing the precision of our knowledge of the initial conditions becomes, at a certain point, unnecessary; the behaviour of the system is well-defined and predictable. Frequently, however, the situation is different. It can happen that the phase space is structured in such a way that in whatever "small" region chosen there are always states present which belong to strongly diverging trajectories.

For these systems (which may be defined as "weakly predictable" or, better, critically dependent upon the initial conditions), the determination of the trajectories would require an infinite precision in the knowledge of the initial conditions.

This singular characteristic of many dynamical systems (already observed by Poincaré for the three-body problem) is at the origin of the unpredictable and "chaotic" behaviour which, surprisingly, may be displayed by the solutions of deterministic differential systems of higher-than-second-order. In particular, there may be "attractors" (that is, regions in the phase space in which, after a sufficiently long time, all the trajectories having certain initial conditions will be confined) which do not have a clearly defined dimension. If one considers, at a certain time, a particular trajectory on the attractor, and the point representing the system over successive periods, it can be seen that it does not return to the previous trajectory, nor approaches it asymptotically. In contrast, after a reasonably long time, the initial state is "forgotten"; consequently, two very similar initial states can give rise to very different trajectories, even though spatially confined on the same attractor.

A "chaotic attractor" (or "strange attractor") thus takes on the appearance of a kind of "hank of string" of trajectories which does not have an integer number of dimensions: it is not one-dimensional (because the trajectory does not return over itself), nor is it two-dimensional (because it is not a surface uniformly covered by the trajectories), nor three-dimensional (because it is not a volume uniformly filled by the trajectories).

The dimensions of a chaotic attractor may be evaluated only by introducing a suitable definition, more wideranging than the usual one. We will not dwell upon this problem, but will only limit ourselves to saying that they have "fractal dimension": a number between n-1 and n.

The possible existence of chaotic attractors allows, at least in principle, the description of "turbulent" behaviours using purely deterministic equations. Interest in this type of description is particularly strong in fluid dynamics and meteorology, but there is also a significant interest, for example, in the study of chemical reactions and lasers (see Chapter 6). Figure 5.19 shows the "Lorenz attractor". This corresponds to a third-order differential system originally obtained in the study of atmospheric turbulence, which (as we shall see) can also describe the behaviour of heavily-pumped lasers.

By means of a parametric analysis, it is possible, in principle, to study not only the bifurcations previously seen, but also the transition towards "deterministic chaos", which corresponds to the birth of a chaotic attractor. This study, however, must be carried out numerically because, for attractors of this type, there are no results of a general character comparable to the Hopf theorem for limit cycles.

The "roads to chaos" travelled by higher-than-second-order differential systems on varying the parameters can be quite different. In many cases, however, there is a cascade of "period

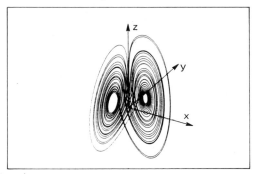

Fig. 5.19 - The Lorenz attractor.

doublings" as a function of a parameter, so that a limit cycle gives rise, by successive bifurcations, to a chaotic attractor. We will examine, in Chapter 6, the results of a numerical simulation of this "road to chaos" for a particular third-order nonlinear system (the Rössler system).

The numerical study of the attractors for a system of differential equations is facilitated by the use of a conceptually simple but extremely powerful instrument: the Poincaré map. The latter can be clearly exemplified in the case of a second-order system. Let us assume that the system is following a particular trajectory and let us study its intersections with a straight line (one of the axes, for example). The $(i + 1)$-th intersection $\xi_{i+1} > 0$ with such an axis is related to the i-th intersection $\xi_i > 0$ by a relationship of the kind

$$\xi_{i+1} = f(\xi_i, P) \tag{5.64}$$

where P is a vector of parameters.

The relationship (5.64) is a "discrete map" whose properties, as we shall see, reflect the trend of the trajectories of the differential equations system. The study of the Poincaré map does clearly not eliminate the necessity of integrating the system over all the time intervals between two successive intersections with the axis being considered. It does allow, however, a projection of the problem onto a one-dimensional manifold, in this way facilitating a graphical analysis (see Fig. 5.20).

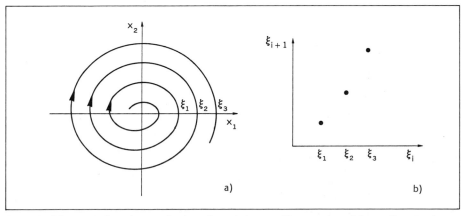

Fig. 5.20 - a) One particular trajectory of a planar dynamical system; b) some points of the one-dimensional map $\xi_{i+1} = f(\xi_i, P)$.

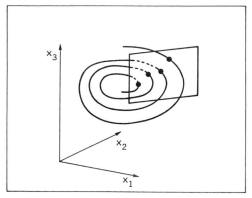

Fig. 5.21 - Poincaré map resulting from the section of a trajectory of a three-dimensional differential system with a fixed plane.

This technique can clearly be extended to the study of higher-order systems. In particular, in the case of third-order systems, the Poincaré map is obtained from the "Poincaré section", i.e. from the intersections of a trajectory with an appropriate surface (often, a plane) and is thus a two-dimensional map (as shown in Fig. 5.21).

For the two-dimensional map corresponding to the three-dimensional dynamical system, a further dimensional reduction may be attempted by projecting it onto a line of the surface under consideration. The resulting one-dimensional map, in many cases, contains the essential information on the initial differential system.

Although this is not always true, since the present work is of an introductory nature we will assume that it is possible in all cases to reduce to one-dimensional maps. We will then examine a particularly simple map which, however, does show many important aspects of the subject matter.

We will thus briefly study the "logistic map"

$$\xi_{i+1} = \alpha \, \xi_i (1 - \xi_i) \tag{5.65}$$

where α is a parameter as a function of which the map itself will be studied. Let us assume that $0 < \xi_i < 1$.

If we require that the map transforms such an interval of values of ξ_i into itself, not all values of α can be allowed, but only values between 0 and 4. If $\alpha < 0$, in fact, positive values of ξ_i will correspond to negative values of ξ_{i+1}; also, if $\alpha > 4$, it is possible to have $\xi_{i+1} > 1$ for some values of $\xi_i < 1$.

The name "logistic map" underlines the fact that Eq. (5.65) is a discrete version of Eq. (5.11). It can be seen, looking at this simple example, that the maps may be of quite general interest also as mathematical models of "naturally" discrete systems. The schematic representation of dynamical systems by means of differential equations does, in fact, implicitly assume that the variables have a physical significance at any point in time. This is not necessarily true: for example, referring back to population dynamics, it may happen that the births take place only in a certain period of the year. In this case, the differential equation may turn out to be an inadequate representation, because it distributes over the whole year events which are concentrated into only a part of the year. Moreover, the numerical solution of a differential equation necessarily leads to a map, so that the study of maps may be of notable interest even for the theory of numerical integration.

Let us now consider the map (5.65), where the relationship between ξ_{i+1} and ξ_i is of the

parabolic type shown in Fig. 5.22a. This Figure suggests the possibility of graphically constructing the successive iterations defined by Eq. (5.65).

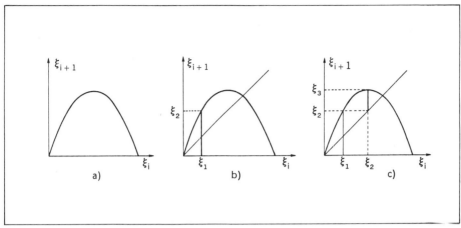

Fig. 5.22 - a) Graphical representation of the parabolic dependence of ξ_{i+1} upon ξ_i expressed by Eq. (5.65);
b) graphical construction of ξ_2 starting from ξ_1; **c)** graphical construction of ξ_3 starting from ξ_2.

For any value of ξ_1 between 0 and 1, Eq. (5.65) gives a corresponding value ξ_2 on the parabola
(Fig. 5.22b). At this point, using the diagonal straight line, the value of ξ_2 is projected on the
horizontal axis, and, using the parabola, the value of ξ_3 is obtained (Fig. 5.22c). The procedure
may be repeated, at this point, indefinitely. It is then interesting to study the asymptotic behaviour of the map, searching, in particular, for possible "attractors", as in the case of differential
equations.

It can be seen, first of all, that the map (5.65) allows "fixed points" at the intersections
of the parabola with the bisector of the first quadrant. The fixed points of the maps are the
discrete equivalents of the equilibrium points of the differential equations: if the initial condition
exactly corresponds to the fixed point, successive iterations will always give the same point. For
the map considered, the fixed points are:

$$\xi^* = 0 \qquad \xi^* = 1 - \frac{1}{\alpha} \tag{5.66}$$

For $0 < \alpha \le 1$ only $\xi^* = 0$ lies between 0 and 1. If $\xi_1 \ne 0$, it can immediately be seen, even
graphically, that $\xi_i \underset{i \to \infty}{\longrightarrow} 0$ and the fixed point $\xi^* = 0$ is thus asymptotically stable. If $1 < \alpha \le 3$,
both points (5.66) are solutions of Eq. (5.65), and there is a change in stability: the fixed point
$\xi^* = 0$ becomes unstable, while the fixed point $\xi^* = 1 - 1/\alpha$ becomes asymptotically stable
(Fig. 5.23a).

The change in stability for the fixed point $\xi^* = 0$ is associated with the fact that the modulus of the derivative of the function $f(\xi, \alpha)$ exceeds 1 at $\xi^* = 0$. In an analogous manner, we can
see that the fixed point $\xi^* = 1 - 1/\alpha$ remains stable only while $|f'(\xi^*, \alpha)| < 1$, after which, for
$\alpha > 3$, it also becomes unstable.

If we consider Eq. (5.65) as a Poincaré map corresponding to a differential system, the
fixed points of the map itself may correspond to equilibrium points, or even to limit cycles. In
particular, in the case of Fig. 5.20, the Poincaré map corresponding to the intersection of the
trajectories with the x_1 axis has an unstable fixed point at the origin. If, on examining higher

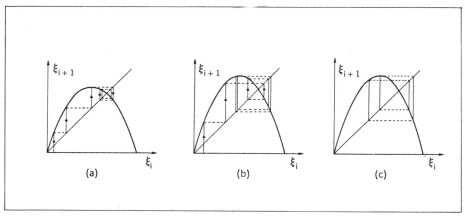

Fig. 5.23 - a) Asymptotically stable fixed point (α between 1 and 3 in Eq. (5.65)); b) asymptotic convergence towards a "hopping" between two values ($3 < \alpha \leqslant 3.4$); c) asymptotic convergence towards a "hopping" between four values ($3.4 < \alpha \leqslant 3.54$).

values of ξ_i, the map were to show also a stable fixed point (as in Fig. 5.23a), this would indicate that the unstable focus of Fig. 5.20 is surrounded by an asymptotically stable limit cycle.

Let us now return to Eq. (5.65), and briefly examine the asymptotic behaviour of the map for $\alpha > 3$. It can be seen graphically (see Fig. 5.23 b) that for values of α only slightly above 3 (between 3 and 3.4, to be precise), there is a convergence towards a kind of "hopping" between two different values, while the fixed point corresponding to the intersection of the parabola $f(\xi, \alpha)$ with the bisector of the first quadrant is, in this case, unstable. Let us once again consider Eq. (5.65) as a Poincaré map for a differential system: as the fixed point indicated the existence of a limit cycle, the "hopping" between two values indicates a doubling of the period of the limit cycle itself. On further increasing α, there is a further doubling of the period: as can be shown numerically, there is "period two" for $3.0 < \alpha \leqslant 3.4$, "period four" for $3.4 < \alpha \leqslant 3.54$, "period eight" for $3.54 < \alpha \leqslant 3.56$, and so on. It can be seen that the values of α for which there is a doubling of the period increase towards a limit value $\alpha_\infty = 3.5699$. It can also be proved that the sequence of the values of α for which there is an n-th doubling of the period is such that

$$\lim_{n \to \infty} \frac{\alpha_n - \alpha_{n-1}}{\alpha_{n+1} - \alpha_n} = \delta$$

where

$$\delta = 4.6692$$

The number δ (Feigenbaum's "universal number") describes the doubling of the period not only for Eq. (5.65), but also for a large class of unimodal maps. At α_∞ the map shows an aperiodic behaviour: this corresponds to the rise of a "chaotic attractor" in the associated differential system. For $\alpha > \alpha_\infty$ the system enters the so called "chaotic region" where the aperiodic behaviour may be interrupted by "windows" of periodicity; in each of these windows, a period doubling as a function of α is also observed.

6

Self-Organisation in Physical and Chemical Systems

6.1 INTRODUCTION

Some examples of self-organisation processes in physical and chemical systems will now be considered. In the preceding chapters, the concepts and the mathematical techniques necessary for such a study have already been outlined. Undoubtedly, the most fascinating aspect of these phenomena is their relative independence of environmental stimuli: i.e., one is dealing with self-organisation.

Mathematically speaking, as we shall see, such phenomena are represented by a change in the stability properties of the equations which regulate the evolution of the order parameters of the system, when there are variations in a control parameter (the level of pumping in a laser, the temperature gradient in Bénard instability, etc.). The control parameter, however, has a very modest information content, and can not prescribe the new state of the system. This new state then arises mainly as an independent response of the system to an environmental change.

When a system is close to instability, fluctuations play a crucial role. As we have, in fact, seen in Chapter 3, in these situations the restoring force, responsible for the damping of the fluctuations around a stationary stable state, tends to zero for at least one of the parameters. The fluctuations, near the critical point, may thus grow to macroscopic dimensions. In some cases the system may evolve beyond the threshold of instability, towards different states which, a priori, are equally, or almost equally, probable.

For example, in the case of the "Bénard rolls", the direction of rotation may be clockwise or counterclockwise. In these cases, the effective evolution of the system is determined by the first fluctuation which makes itself felt close to the critical point. If one considers a large number of copies of the system, an equal distribution between the two directions of rotation will be observed, but in any particular case of interest it will be seen that the symmetry will be destroyed by the dominant fluctuation.

At the end of this chapter, another example of self-organisation will be examined, differing from the cases examined previously in that the variations in one of the control parameters are random. This is the case in which a parameter of the evolution equation of the system is given by the sum of a deterministic term, equal to its average value, and a fluctuating force of a white noise type.

The phenomenon of so-called multiplicative noise will be met once again. As we shall see, the system considered in this example can respond to the fluctuations in the control parameter, reorganising itself, when the intensities of such fluctuations exceed a critical value. These peculiar phenomena of self-organisation, known as noise-induced transitions, can not be studied using bifurcation analysis of the associated deterministic system, but only by using stochastic methods, and, in particular, by analysing the nature of the stationary solutions of the Fokker-Planck equation of the system.

6.2 PRINCIPLES OF LASER OPERATION

The laser (the word derives from the initial letters of "Light Amplification by Stimulated Emission of Radiation") is a device which produces coherent and monochromatic radiation through the suitable use of "stimulated emission", a typically quantum effect. "Stimulated emission" refers to the incidence of electromagnetic radiation of a suitable frequency upon a quantum system in an excited state which may induce a transition to a lower energy state accompanied by the emission of radiation in phase with the incident radiation and of the same frequency.

As will become clear later, the laser is a system which is far from equilibrium (because it requires the presence of external excitation) whose operation critically depends upon a cooperative process involving a large number of "active elements". The laser effect may occur between two energy levels of an atom, of an ion, of a molecule (not only between electronic levels, but also between vibrational and rotational levels), or of an entire crystal (the latter is the case of semiconductors).

"Active materials" may be in the gaseous, liquid or solid state. The study presented below is particularly suitable for the description of a solid state laser in which the "active atoms" are considered to be in fixed positions within the host material. The analysis carried out will thus be quite simple but, at the same time, will also yield results of a general validity.

It is worth recalling, at this point, the concept of energy levels for an atomic system. Quantum mechanics requires that an electron bound to an atom can take only a well-defined series of energy values. In other words, the electron is to be found only at certain energy levels (which in solids and liquids may group themselves into allowed energy "bands").

For simplicity, only two possible energy levels of the outer electrons of the active atoms will be considered (a "ground state" and an "excited state"). At thermal equilibrium, the statistical distribution will be characterised by a greater population in the lower energy state than in the higher energy one. As we shall see, external excitation of a suitable frequency may, under certain conditions, cause an "inversion" of the populations of the two levels.

In the presence of a weak external excitation (and, therefore, of limited inversion) the laser acts as a normal lamp, in the sense that every atom, after being excited, decays spontaneously with the emission of a photon in a completely uncorrelated manner with respect to the other atoms. If, however, the excitation is sufficiently intense, the photons produced (initially by spontaneous emission) may significantly alter the electric field present in the available volume which, in its turn, by a feedback process, gives rise to stimulated emission.

The stimulated emission (made possible by the inversion produced by external excitation) then prevails over the spontaneous emission, and a highly ordered situation arises, even though the individual electrons were excited by completely uncorrelated external pumping.

To favour the establishment of a certain dynamic order, a particular geometry of the laser device is necessary, as shown schematically in Fig. 6.1. Two mirrors act as the axial boundary surfaces, at least one of these being a half-mirror through which the light beam can leave the cavity containing the active material.

This configuration allows the selection of the frequency and direction of the "modes" of the electromagnetic field. A "mode" is intended as a stationary spatial configuration of the field within the "cavity" formed by the mirrors. For such a configuration, the maximum amplitude

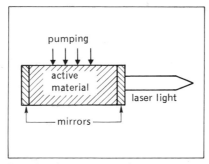

Fig. 6.1 - Schematic picture of a solid-state laser.

of oscillation for each frequency at a given point is constant with respect to time.

In order to construct an approximate but simple mathematical model for the laser it is necessary to individuate some "order parameters" which allow a macroscopic description of the system consisting of an extremely large number of microscopic elements.

Let us choose, for this description, two parameters which are intuitively very important: the number n of coherent photons present in the cavity and the population inversion D, that is, the difference between the number of electrons excited to the higher level (N_2) and the number of electrons in the lower level (N_1).

As we have already mentioned, the mirrors at the ends of the cavity serve the function of selecting, from among the electric field "modes", those characterised by specific frequencies (in particular, in a one-dimensional model, the length of the cavity must be equal to an integral number of half wavelengths) and by well-defined stationary spatial configurations.

As we shall see below, on increasing the external excitation the system becomes unstable, and there is a sharp rise in the number of coherent photons corresponding to a particular field mode. Later, we shall mention the problem of the "competition" between the different modes which occurs in this phase of the process.

Here we will only report the result, according to which only one mode prevails, so that the strength of the corresponding electric field (or equivalently, the number of coherent photons) takes on the role of "order parameter" for the entire system. We will therefore consider here only coherent photons having the same frequency.

We will also treat the number n of photons present as a continuous variable in time which is independent of the spatial coordinates.

This latter hypothesis clearly corresponds to a spatial averaging operation.

The variation of n not only depends upon the emission and absorption processes, but also upon the losses due to the mirrors, absorption by the lateral walls of the cavity, etc. Let us assume, for simplicity's sake, that the loss of photons per unit time is proportional to the number of photons within the cavity through a coefficient $1/\tau$.

We will use N_1 and N_2 for the total numbers of electrons to be found, respectively, in levels 1 and 2. The number of photons generated per unit time by stimulated emission may be assumed to be proportional to the product of n and N_2, and can therefore be written as WnN_2 (where W is a suitable coefficient).

The number of photons absorbed in unit time is WnN_1. Simple thermodynamic considerations show that the coefficient W must be the same for these two processes. In the following, we will assume that W is independent of the spatial coordinates.

We thus obtain, for the number of coherent photons present in the cavity, the equation

$$dn/dt = -n/\tau + Wn(N_2 - N_1) \tag{6.1}$$

In our study of the laser effect, we have avoided any consideration of spontaneous emission in the equation for the number of coherent photons, because the latter process gives rise to mutually uncorrelated photons. This consideration, however, suggests the possibility of treating the spontaneous emission as a noise term in an otherwise deterministic equation.

Let us now study the population inversion $D = N_2 - N_1$. This may be modified following external pumping, spontaneous emission, non-radiative decay, photon absorption, and finally stimulated emission.

Let W_{21} (equal for all the active atoms) be the probability of decay per unit time of an electron from level 2 to level 1 by spontaneous emission and non-radiative decay, and let W_{12} be the probability of excitation from level 1 to level 2. Under conditions of thermal equilibrium and in the absence of external "pumping", the coefficient W_{12} only takes into account the thermal excitation and has a very low value (level 1, as we have already mentioned, is much more densely populated than level 2).

On the other hand, not even the presence of external radiation which directly promotes electrons from level 1 to level 2 can bring about a population "inversion" between the levels. In fact, when N_2 becomes equal to N_1, the stimulated emission per unit time is equal to the absorption, and the material becomes transparent to the pumping.

An indirect process is then necessary for modifying the value of the coefficient W_{12}, in order to circumvent the above limitation. The simplest way of pumping in order to produce an inversion is by promoting the electrons from level 1 to a third energy level which is higher than level 2, characterised, however, by a rapid spontaneous decay to level 2. In this way, it is thus possible to consider W_{12} modified and simultaneously maintain the sum of N_1 and N_2 constant.

Excitation due to the absorption of the photons present in the cavity may be represented by the term $- WnN_1$ in the equation for N_2. By analogy, the stimulated emission gives rise to a term WnN_2 in the equation for N_1 and to the same term, of opposite sign, in the other equation. We thus have

$$\frac{dN_2}{dt} = -N_2 W_{21} + N_1 W_{12} - Wn(N_2 - N_1)$$

$$\frac{dN_1}{dt} = N_2 W_{21} - N_1 W_{12} + Wn(N_2 - N_1)$$

(6.2)

By adding and subtracting $N_1 W_{21}$ to the r.h.s. of the first equation and $N_2 W_{12}$ to the r.h.s. of the second equation, this system may be rewritten as

$$\frac{dN_2}{dt} = -(N_2 + N_1)\, W_{21} + N_1\, (W_{12} + W_{21}) - Wn(N_2 - N_1)$$

$$\frac{dN_1}{dt} = -(N_2 + N_1)\, W_{12} + N_2\, (W_{12} + W_{21}) + Wn(N_2 - N_1)$$

Subtracting the second equation from the first, we obtain

$$\frac{dD}{dt} = (N_2 + N_1)\, (W_{12} - W_{21}) -$$
$$- D\, (W_{21} + W_{12}) - 2Wn(N_2 - N_1)$$

Setting

$$N_2 + N_1 = N$$

(6.3)

$$D_{p0} = \frac{W_{12} - W_{21}}{W_{12} + W_{21}} N \tag{6.3}$$

$$\frac{1}{T} = W_{12} + W_{21}$$

The resulting equation for D may be rewritten as

$$\frac{dD}{dt} = \frac{D_{p0} - D}{T} - 2WnD \tag{6.4}$$

where both D_{p0} and T depend upon the intensity of the external "pumping" through the coefficient W_{12}. This coefficient, in its turn, may be considered as directly proportional to the electric power used for the pumping. We have thus arrived, with very simple considerations, at a system of first-order nonlinear differential equations

$$\frac{dn}{dt} = -\frac{n}{\tau} + WnD$$
$$\frac{dD}{dt} = \frac{D_{p0} - D}{T} \quad 2WnD \tag{6.5}$$

Formally, we can consider this system as a variant of the Lotka-Volterra system examined in Chapter 5, where the "prey" (represented here by the variable D) undergoes a spontaneous decay with a time constant T and is supplied from the exterior at a constant rate D_{p0}/T. As we can immediately see, the system allows two points of equilibrium:

$$(n_0, D_0) = (0, D_{p0})$$

$$(n_0, D_0) = \left(\frac{D_{p0} W\tau - 1}{2TW}, \frac{1}{\tau W} \right)$$

Recalling that n represents the number of photons, we can say that the latter point of equilibrium only has physical significance for $D_{p0} > 1/\tau W$.

Let us first of all study the stability of the point $(0, D_{p0})$. Linearising the differential equations around this equilibrium configuration, we have:

$$\frac{dv}{dt} = \left(D_{p0} W - \frac{1}{\tau} \right) v$$
$$\frac{d\delta}{dt} = -\frac{\delta}{T} - 2WD_{p0} v \tag{6.6}$$

The Jacobian matrix in this case allows the following eigenvalues:

$$\lambda_1 = \frac{D_{p0} W\tau - 1}{\tau}$$

$$\lambda_2 = -\frac{1}{T}$$

It can be seen that λ_2 is always negative, while the sign of λ_1 depends upon the value of $D_{p_0} W\tau$ and thus on the intensity of the pumping. In particular, the point of equilibrium $(0, D_{p_0})$ is a stable node when $D_{p_0} W\tau < 1$.

From the physical point of view, this means that for low values of D_{p_0} (limited pumping) the laser behaves as a normal lamp, i.e., the emitted photons are incoherent, and therefore $n = 0$.

If, however, $D_{p_0} > 1/\tau W$, the equilibrium point becomes a saddle point (unstable). The instability arieses under the same condition which allows the existence of the new, physically significant, equilibrium configuration

$$\left(\frac{D_{p_0} W\tau - 1}{2TW}, \ \frac{1}{\tau W} \right)$$

the stability of which we will now consider.

Linearising the equations about the equilibrium configuration, we have

$$\frac{d\nu}{dt} = \frac{D_{p_0} W \tau - 1}{2T} \delta$$

$$\frac{d\delta}{dt} = -\frac{D_{p_0} W \tau - 1}{T} \delta - \frac{2}{\tau} \nu$$

$$(6.7)$$

In this case, the eigenvalues of the Jacobian matrix are given by the characteristic equation:

$$\lambda^2 + \frac{D_{p_0} W\tau}{T} \lambda + \frac{D_{p_0} W\tau - 1}{T\tau} = 0$$

On examining the discriminant of this equation, it can be seen that the algebraic sign of the eigenvalues at this equilibrium point depends upon two control parameters: T/τ and, say, $D_{p_0} W\tau$. In any case, when $D_{p_0} W\tau > 1$, the equilibrium point

$$\left(\frac{D_{p_0} W\tau - 1}{2TW}, \ \frac{1}{\tau W} \right)$$

which becomes physically significant, turns out to be stable. In particular, when $T < \tau$, (a particularly interesting physical case) the equilibrium point becomes a stable node if $D_{p_0} W\tau > 1$ and an unstable saddle if $D_{p_0} W\tau < 1$. When $D_{p_0} W\tau = 1$ there is a bifurcation with a change of stability.

We have thus identified a critical value of the pumping parameter, $D_{p_0} = 1/\tau W$, which constitutes the threshold of laser activity. This depends, as intuition would tell us, not only upon the losses (combined together in the time constant τ), but also upon the efficiency of the stimulated emission (expressed by the coefficient W).

Noting that the pumping parameter D_{p_0} is also proportional to the total number of active atoms, an unusual fact presents itself: the transition from "lamp" to "laser" operation mode may be obtained, at least in principle, even under constant pumping power by increasing the number of active atoms. A simple increase in the number of components in the system (all of which are equal) would thus give rise to a profound qualitative change in the global behaviour of the system itself.

Let us now consider the case where the intrinsic dynamics of the inversion is much more rapid than those of the photons present in the cavity ($T \ll \tau$). This condition, occurring frequently in some types of laser, may also be reached, at least in principle, by means of sufficiently intense pumping as can be seen from Eq. (6.3) for $1/T$.

By introducing the "direct adiabatic approximation" $dD/dt \sim 0$ into the previous system of differential equations, we obtain

$$\frac{dn}{dt} = -\frac{n}{\tau} + \frac{WD_{p_0}n}{1 + 2TWn} \tag{6.8}$$

This equation clearly allows the same equilibrium values for n found previously, i.e.

$$n_0 = 0$$

$$n_0 = \frac{D_{p_0}W\tau - 1}{2TW}$$

We can begin the study of the stability (including the "global" stability) by noting that the "generalised force" acting upon n may be considered as deriving from the potential

$$V(n) = \frac{n^2}{2\tau} - D_{p_0}W\left[\frac{n}{2TW} - \frac{1}{(2TW)^2}\ln(1 + 2TWn)\right] \tag{6.9}$$

The second derivative of this potential with respect to n is

$$\frac{\partial^2 V}{\partial n^2} = \frac{1}{\tau} - \frac{D_{p_0}W}{1 + 2TWn} + \frac{2TD_{p_0}W^2 n}{(1 + 2TWn)^2}$$

and, at the points of equilibrium, we have

$$\left.\frac{\partial^2 V}{\partial n^2}\right|_{n=0} = \frac{1}{\tau} - WD_{p_0}$$

$$\left.\frac{\partial^2 V}{\partial n^2}\right|_{n=\frac{D_{p_0}W\tau - 1}{2TW}} = \frac{D_{p_0}W\tau - 1}{D_{p_0}W\tau^2}$$

From this, we can again see that the non-zero equilibrium point becomes stable when it takes on a physical significance, that is, above the laser threshold. In a corresponding manner, the point $n_0 = 0$ becomes unstable. The qualitative behaviour of the potential as a function of D_{p_0} is shown in Fig. 6.2: with increasing D_{p_0}, there is a "continuous transition" from curve (a) to curve (b).

With different experimental configurations, the transition undergoes a drastic change and may take on the aspect of a "discontinuous transition" or "catastrophe".

Let us now suppose that a "saturable absorber" is interposed between the active material and the half-reflecting mirror. This absorber has the following properties: if it is irradiated with a weak light source, it shows a certain absorption; if the intensity increases, it becomes increasingly transparent. We can take the presence of this absorber into account by introducing, into the equation for n, a decay time τ which depends upon the number of photons present according to the equation

$$\frac{1}{\tau} = \frac{1}{\tau_0}\left[1 + \frac{K_s}{1 + n/n_s}\right] \tag{6.10}$$

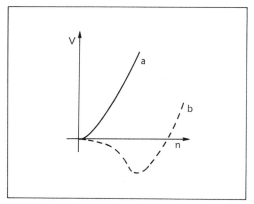

Fig. 6.2 - Qualitative behaviour of the potential of Eq. (6.9) associated to the laser transition in the adiabatic approximation (curves (a) and (b) refer respectively to a lower and a higher value of the parameter D_{p_0}).

The potential $V(n)$ then takes the form

$$V(n) = \frac{n^2}{2\tau_0} + \frac{K_s}{\tau_0}\left[n_s n - n_s^2 \ln\left(1 + \frac{n}{n_s}\right)\right] -$$
$$- D_{p_0} W\left[\frac{n}{2TW} - \frac{1}{(2TW)^2}\ln\left(1 + 2TWn\right)\right]$$

(6.11)

Figure 6.3 shows the qualitative behaviour of the potential as a function of n for different values of the parameter D_{p_0}. It can be seen that for fairly high values of D_{p_0} there is the possibility of a "catastrophe", with the corresponding hysteresis effect.

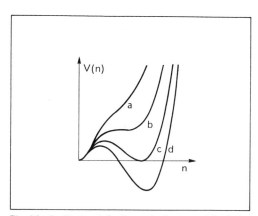

Fig. 6.3 - Qualitative behaviour of the potential of Eq. (6.11) in the presence of a saturable absorber for different values of the parameter D_{p_0}. Curves (a) to (d) refer to increasing values of D_{p_0}.

For the sake of simplicity we have considered so far only one "mode" of the field engendered by the coherent photons within the cavity. If we now associate a decay time τ and a coefficient of stimulated emission W to each mode, we can write the following equations for two modes which are fairly close to resonance:

$$\frac{dn_1}{dt} = -\frac{n_1}{\tau_1} + W_1 n_1 D$$

$$\frac{dn_2}{dt} = -\frac{n_2}{\tau_2} + W_2 n_2 D \qquad (6.12)$$

$$\frac{dD}{dt} = \frac{D_{p_0} - D}{T} - 2D\,[W_1 n_1 + W_2 n_2]$$

Formally, we can say that these equations describe the "competition" between two different entities, n_1 and n_2, for a "resource" D which enters the system at a constant rate. Without considering the problem of solving these equations, we can see that, under stationary conditions, the two modes can not be present simultaneously. This would, in fact, require, at one and the same time, that

$$D = \frac{1}{\tau_1 W_1} \qquad \text{and} \qquad D = \frac{1}{\tau_2 W_2}$$

whereas these values are generally different. The result obtained from a more complete study of the system of equations is very simple: only the mode which involves lower losses (large τ) and which is closer to the resonance (large W) survives.

So far, we have adopted a purely deterministic treatment. It is clear, however, that Eq. (6.8) obtained for n by the adiabatic approximation may be transformed into a Langevin equation by adding a suitable stochastic term.

As we know, it is the noise of a system which allows the "exploration" of states close to the equilibrium one, and therefore forces the system to make transitions according to the variations in the parameters (in our case, in the parameter D_{p_0}).

It is also possible to write a Fokker-Planck equation corresponding to the Langevin equation obtained from Eq. (6.8) by adding a noise term. To this end, however, it is necessary to know the diffusion coefficient, or, in other words, the self-correlation function of the stochastic "force".

In our case, the determination of the diffusion coefficient would require complicated quantum mechanical considerations. We will thus only approximately evaluate the contribution coming from the spontaneous emission and from the excitation deriving from thermally generated photons, ignoring the effect of the mirrors and walls of the cavity.

For this purpose, we note that every one of the elementary events considered gives rise to a unit contribution (positive or negative) to the number of photons, whereas the average time between two events is equal to $1/W$ for a single atom and to $1/WN$ for all the active atoms. In the case of independent events, it can easily be demonstrated that the diffusion coefficient (for which we will use the letter C in order to avoid confusion with the parameter D) is equal to

$$C = F_0^2 / \tau$$

where F_0 is the intensity of the stochastic "force" and τ the average time between two events. In our case, we then have $C = WN$. Supposing that even in a more accurate evaluation the coefficient C turns out to be independent of n, we can then write the Fokker-Planck equation in the form

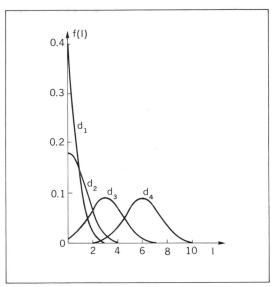

Fig. 6.4 - **Stationary distribution function of the normalised electric field intensity as a function of the pumping parameter d. The value d_2 corresponds to the inversion threshold.**

$$\frac{\partial f}{\partial t} = \frac{\partial}{\partial n}\left[K(n)f + \frac{1}{2}C\frac{\partial f}{\partial n} \right] \tag{6.13}$$

with

$$K(n) = -\frac{n}{\tau} + \frac{D_{p_0}Wn}{1+2TWn} \tag{6.14}$$

The stationary solution of the Fokker-Planck equation can be obtained in the form

$$f = f_0 e^{\frac{2V(n)}{C}} \tag{6.15}$$

where f_0 is a suitable normalisation factor and $V(n)$ is the potential of Eq. (6.9). A knowledge of the distribution function $f(n)$ allows one to predict the magnitude of the fluctuations in the number of coherent photons (and therefore in the intensity of the electric field) as a function of the pumping. Figure 6.4 shows the stationary distribution of the (normalised) intensity of the electric field, I, as a function of a suitable pumping parameter d, which is a function of D_{p_0}. The value d_2 corresponds to the inversion threshold, which marks the transition towards the dissipative stationary state corresponding to the laser effect.

6.3 A MORE SATISFACTORY SEMICLASSICAL TREATMENT OF THE LASER EFFECT

The simple treatment of the laser developed in Section 6.2 is clearly not capable of properly describing all the different physical situations. Among them it is worth mentioning the unusual experimental behaviour of certain types of highly pumped lasers which may exhibit a transition from a continuous operation to a self-pulsing one. Under certain conditions, the emission can also become "turbulent", that is, chaotic and unpredictable.

On the other hand, it can be shown that the system represented by Eq. (6.5) does not allow limit cycles. This statement, which can not be further justified here, may be considered as acceptable due to the absence of Hopf bifurcations on varying the pumping parameter.

The possibility of a self-pulsing operation of lasers (corresponding to a limit cycle) suggests that the system (6.5) is not sufficient for the description of laser dynamics, and that it is thus necessary to consider another order parameter besides D and n.

Under normal conditions, the latter parameter is ignored because it can be adiabatically eliminated, whereas under different conditions it may become significant. As we shall see, this extra variable describes the coupling between the inversion parameter and the electric field. So far, in fact, it has been implicitly assumed that the decay process of the electron is instantaneously followed by emission, and the absorption by excitation. If this is no longer true, the system may be described by three first-order differential equations in place of two.

A rigorous study of the physics of the laser may only be developed within the framework of quantum electrodynamics, because the fundamental processes involved (and, above all, stimulated emission) can not be described within classical physics. For our purposes, however, a semiclassical treatment is sufficient, in which the electric field is treated classically by means of the Maxwell equations, but with the introduction (in a more or less heuristic manner) of hypotheses for the description of the behaviour of the electrons, which are theoretically justifiable only within a quantum context.

An acceptable description of the electrons is provided by Thompson's model, which refers to a simple elastic bond. Within the framework of this model, the emission can be considered as the irradiation from an electric dipole (a process which thus requires a finite time), while the absorption is represented by the initiation of oscillations of the dipole itself. Each dipole is associated with a polarisation

$$\vec{p}_i = q(\vec{x}_i - \vec{x}_{i_0})$$

where q is the electronic charge and the vector $\vec{x}_i - \vec{x}_{i_0}$ identifies the deviation of the i-th oscillator from its equilibrium position.

If, for simplicity, we assume that only the active atoms contribute to the polarisation, then the total atomic polarisation may be expressed as

$$\vec{P} = \sum_i \vec{p}_i$$

where the summation is extended over all the active atoms.

The electric field is obtained from the Maxwell equations, and in particular from the relationship

$$\vec{\nabla} \times \vec{H} = \frac{4\pi}{c} \vec{J} + \frac{1}{c} \frac{d}{dt}(\vec{E} + 4\pi\vec{P}) \tag{6.16}$$

where

$$\vec{J} = \sigma\vec{E} \qquad \vec{H} = \vec{\nabla} \times \vec{A} \qquad \vec{E} = -\frac{1}{c}\frac{d\vec{A}}{dt} \tag{6.17}$$

Adopting the Coulomb gauge for the vector potential $\vec{A}$

$$\vec{\nabla} \cdot \vec{A} = 0$$

and recalling that

$$\vec{\nabla} \times (\vec{\nabla} \times \vec{A}) = \vec{\nabla}(\vec{\nabla} \cdot \vec{A}) - \nabla^2 \vec{A}$$

we obtain

$$-\nabla^2 \vec{A} + \frac{1}{c^2} \ddot{\vec{A}} + \frac{4\pi\sigma}{c^2} \dot{\vec{A}} = \frac{4\pi}{c} \dot{\vec{P}} \qquad (6.18)$$

from which, on differentiating with respect to time and expressing $\vec{A}$ as a function of $\vec{E}$, we obtain

$$-\nabla^2 \vec{E} + \frac{1}{c^2} \ddot{\vec{E}} + \frac{4\pi\sigma}{c^2} \dot{\vec{E}} = -\frac{4\pi}{c^2} \ddot{\vec{P}} \qquad (6.19)$$

This equation expresses the dynamic effect of the polarisation on the electric field and therefore, within the framework of Thompson's model, the effect on the electric field of the emission and the absorption of photons by the active atoms.

On the other hand, the polarisation is influenced, in its turn, by the electric field via the absorption and the stimulated emission processes. Each one of the active atoms may be considered as a damped (due to its irradiation) and forced (due to the absorption and the stimulated emission) oscillator.

In the hypothesis that all the active atoms have the same angular frequency of oscillation (corresponding to the central frequency of the emission spectrum), we can write

$$\ddot{\vec{p}}_i + \gamma\dot{\vec{p}}_i + \omega_o^2 \vec{p}_i = \vec{F}(\vec{x}_i, t) \qquad (6.20)$$

The coefficient γ expresses the energy loss through irradiation, and determines the width of the emission spectrum.

The expression for the forcing term $\vec{F}$ as a function of the electric field can not be obtained on the basis of purely classical considerations. Consequently, here we will only present plausible arguments for an expression which could be adequately justified only in a quantum context.

In order to simplify our discussion, we will consider the one-dimensional case; the vectors considered here are therefore mutually parallel and the arrows denoting vectorial variables may be omitted in the following.

To develop a plausible physical hypothesis on the form of the forcing term $F(x_i, t)$ we recall that the electric field may prevalently give rise to excitation or to stimulated emission according to the sign of the population inversion. It is thus reasonable to suppose that the forcing term is proportional to E when $D_i < 0$ and to $-E$ when $D_i > 0$; for F, we can then assume an expression of the type

$$F = -KE(x_i, t)D_i \qquad (6.21)$$

The multiplicative coefficient may only be determined on the basis of quantum considerations. We limit ourselves to conjecture (as will be confirmed later) that K must depend upon the transition probability W of the electrons for stimulated emission (or, equivalently, for absorption).

The equation for the polarisation of a single atom thus takes the form

$$\ddot{p}_i + \gamma\dot{p}_i + \omega_o^2 p_i = -K E(x_i, t) D_i \qquad (6.22)$$

Since in the equation for the electric field the sum of the contributions of the polarisation of the single atoms appears, it is worth writing a single equation for the atomic polarisation of the active medium

$$P(x, t) = \sum_i p_i$$

From Eq. (6.22) it follows that

$$\ddot{P}(x, t) + \gamma \dot{P}(x, t) + \omega_0^2 P(x, t) = - KE(x, t) D(x, t) \tag{6.23}$$

where

$$D(x, t) = \sum_i D_i$$

This operation of reducing the variables is particularly simple for solid state lasers, where the active atoms occupy fixed positions. In the case of gas lasers, for example, it would be necessary to proceed with more caution.

The last equation to be considered is the one for the population inversion. Considering at first the processes which do not depend upon the electric field, we can write

$$\frac{dD}{dt} = \frac{D_{p0} - D}{T} \tag{6.24}$$

where both D_{p0} and T depend upon the external pumping.

We must now look for an expression for the dependence of D upon the electric field $E(x, t)$ following the processes of absorption and of stimulated emission.

Let us consider, for this aim, the work carried out in unit time by the electric field on the electrons which undergo energy transitions between the two levels. For this amount of work we will assume the classical expression $E\dot{P}$. Introducing the energy associated with the inversion, $\hbar\omega_o D$, we note that the definition itself of D implies that the effect of this work on the energy associated with the inversion process is multiplied by 2, i.e.

$$\frac{d}{dt}(\hbar\omega_o D) = 2E\dot{P} \tag{6.25}$$

This relationship expresses the fact that it is the coupling between the field and the polarisation which gives rise to the absorption, or to the stimulated emission.

By adding the two contributions, we obtain an equation for the temporal evolution of D

$$\frac{dD}{dt} = \frac{D_{p0} - D}{T} + \frac{2}{\hbar\omega_o} E\dot{P} \tag{6.26}$$

As we already know, the term D_{p0} is associated with the external pumping (it is the latter, in fact, which determines the difference $W_{12} - W_{21}$) and it is also proportional to the total number of electrons available for the transition, and therefore to the total number of active atoms.

As already stressed in Section 6.2, the presence of a cavity limited by mirrors serves to select stationary "modes" of the electric field with particular frequencies and spatial characteristics.

Without entering into a detailed study, we can say that there is a stationary mode for the cavity being considered when, at a suitable frequency, the electromagnetic waves can propagate forwards and backwards without destructive interference, being only attenuated. In the following, we will assume that the presence of such attenuation (including that due to the semitransparency of the mirrors) can be represented by the conductance which appears in the equation for the electric field.

As we know, on increasing the pumping, the system becomes unstable and, at a certain point, a bifurcation occurs followed by a rapid growth of the electric field strength. Referring now to the competition which occurs between different modes during this stage, we recall, from the previous simplified treatment, that only one mode prevails, so that its amplitude takes on the role of an order parameter for the entire system. We will confine ourselves to considering waves of a single frequency, which we assume to be exactly in resonance with the oscillation frequency of the electric dipoles. Considering only progressive waves, for E and for P we can take the following expressions

$$E(x, t) = E^+(t)\, e^{i(K_o x - \omega_o t)} + E^-(t)\, e^{-i(K_o x - \omega_o t)}$$

$$P(x, t) = P^+(t)\, e^{i(K_o x - \omega_o t)} + P^-(t)\, e^{-i(K_o x - \omega_o t)}$$

where $K_o = \omega_o/c$.

Restricting our attention to progressive (or regressive) waves corresponds to a schematisation of the laser as a circular structure, which, therefore, does not possess the reflective boundaries provided by the two mirrors. Naturally, were this hypothesis exactly verified, then it would not be possible to extract radiation from the system.

We also assume that the amplitudes E^+, E^-, P^+, P^- can be considered as time functions which vary very slowly compared to the typical times of the exponential factors. In other words, we assume that two very different time scales can be distinguished, confining our study to what happens over the longer periods of time. We also note that the preceding expressions for E and P contain the further hypothesis that the amplitudes are spatially independent, and therefore the spatial dependence of E and P is contained only in the exponential factors.

Without discussing here the limits of the validity of these hypotheses, let us verify that the resulting model (defined as the "reduced model" in the literature) is capable of providing a qualitative treatment of the different operation regimes of the laser.

With the above hypotheses, we can approximately write

$$\frac{\partial E}{\partial t} \simeq -i\,\omega_o E^+ e^{i\omega_o[(x/c) - t]} + i\,\omega_o E^- e^{-i\omega_o[(x/c) - t]}$$

$$\frac{\partial^2 E}{\partial t^2} \simeq -2i\omega_o \frac{\partial E^+}{\partial t} e^{i\omega_o[(x/c) - t]} +$$

$$+ 2i\omega_o \frac{\partial E^-}{\partial t} e^{-i\omega_o[(x/c) - t]} +$$

$$+ \omega_o^2\,[E^+ e^{i\omega_o[(x/c) - t]} + E^- e^{-i\omega_o[(x/c) - t]}]$$

$$\frac{\partial^2 E}{\partial x^2} \simeq -\frac{\omega_o^2}{c^2}\,[E^+ e^{i\omega_o[(x/c) - t]} + E^- e^{-i\omega_o[(x/c) - t]}]$$

and analogous expressions for P.

Substituting in the equation for the electric field, we obtain for the coefficients of the two exponentials

$$\frac{d}{dt} E^+(t) + 2\pi\sigma E^+ = 2\pi i \omega_o P^+ \tag{6.27}$$

$$\frac{d}{dt} E^-(t) + 2\pi\sigma E^- = -2\pi i \omega_o P^- \tag{6.28}$$

In an analogous manner, we obtain

$$\frac{d}{dt} P^+(t) + \gamma P^+ = -\frac{iK}{2} E^+ D \tag{6.29}$$

$$\frac{d}{dt} P^-(t) + \gamma P^- = \frac{iK}{2} E^- D \tag{6.30}$$

The equation for the inversion parameter then becomes

$$\frac{d}{dt} D(t) = \frac{D_{pu} - D}{T} + \frac{2i}{\hbar} [E^+ P^- - E^- P^+] \tag{6.31}$$

It is interesting to note that the equations obtained previously for the number of photons n and for the inversion parameter D can be derived from these if it is assumed that the polarization "adiabatically" follows the electric field ($\gamma \gg 2\pi\sigma$) and the inversion parameter ($\gamma \gg 1/T$).
From Eqs. (6.29) and (6.30) we obtain

$$P^+ = -\frac{iK}{2\gamma} E^+ D, \qquad P^- = \frac{iK}{2\gamma} E^- D$$

Substituting in Eqs. (6.27) and (6.31), we have

$$\frac{d}{dt} E^+ = -2\pi\sigma E^+ + \frac{\pi\omega_o K}{\gamma} E^+ D \tag{6.32}$$

$$\frac{d}{dt} D = \frac{D_{po} - D}{T} - 2\frac{KD}{\hbar\gamma} E^+ E^- \tag{6.33}$$

We now see that Eq. (6.32) for E^+ is linear with real coefficients, so that it can be divided into two equations, one for the real part A and one for the imaginary part B of $E^+ = A + iB$. Multiplying the first equation by A, the second by B and adding, and indicating the field intensity with $I = A^2 + B^2$, we obtain

$$\frac{dI}{dt} = -4\pi\sigma I + \frac{2\pi\omega_o K}{\gamma} I D$$

$$\frac{dD}{dt} = \frac{D_{po} - D}{T} - \frac{2K}{\hbar\gamma} I D \tag{6.34}$$

Formally, these equations coincide with Eqs. (6.5) if one notes that the number of photons present is proportional to I. It can also be seen that σ is inversely proportional to τ and, as was anticipated, the coefficient K is proportional to the transition probability W.

On increasing the pumping, the inversion relaxation time continually decreases, until it is no longer possible to ignore the polarisation dynamics, considering it as a "slave" variable, as done in Eqs. (6.34). It then becomes necessary to resort to the system of "reduced" equations (6.27) - (6.31) in its general form.

In order to further simplify Eqs. (6.27) - (6.31), let us assume that the phase shift between E and P is constant (it can be verified numerically that this is a generally acceptable hypothesis). On this point we note that at equilibrium we have

$$E^+/P^+ = i\omega_o/\sigma$$

This suggests that we choose E^+ as being real and P^+ imaginary. Setting

$$E^+ = x, \ E^- = x, \ P^+ = -iy, \ P^- = iy \ \text{and} \ D = z$$

we obtain the system

$$\dot{x} = -2\pi\sigma x + 2\pi\omega_o y$$

$$\dot{y} = \frac{K}{2}xz - \gamma y \qquad\qquad\qquad (6.35)$$

$$\dot{z} = -\frac{4}{h}xy + (D_{p_0} - z)$$

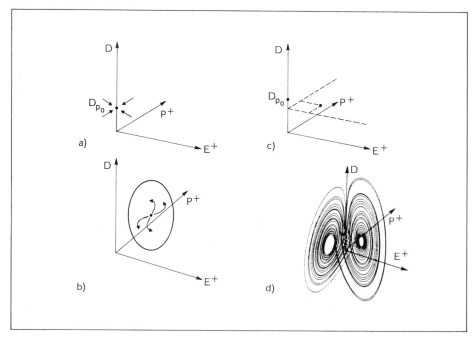

Fig. 6.5 - Schematic illustration, in the space spanned by the reduced variables E^+, P^+ and D, of the different operation regimes of the laser suggested by Eqs. (6.35): (a) lamp mode; (b) continuous laser mode; (c) pulsed laser mode; (d) turbulent laser mode.

As can be seen, these equations formally correspond to the Lorenz system mentioned in Chapter 5.

A numerical study of these equations shows that, besides the laser transition (also described by the simple equations (6.34) for n and D), a second transition takes place for higher values of D_{p_0}.

In three-dimensional space (E^+, P^+, D) the laser transition corresponds to a transition from the situation shown in Fig. 6.5 a to that in Fig. 6.5 b. On further increasing D_{p_0}, the equilibrium point corresponding to the normal operation of the laser becomes unstable, and a Hopf bifurcation occurs with the growth of a limit cycle shown in Fig. 6.5 c (pulsed operation).

Furthermore, for particular values of the parameters σ, γ and T, Eqs. (6.35) suggest the possible existence of a "turbulent" regime for the laser: correspondingly, the limit cycle could transform itself into a "strange attractor" of the type shown in Fig. 6.5 d.

We will not discuss here the problems posed by the experimental investigation of this behaviour of highly-pumped lasers. The interested reader is referred to the bibliographical notes relating to this Chapter.

Is should be noted, however, that a turbulent behaviour such as that predicted by the system (6.35) has not been observed experimentally, due to the impossibility of exploring some regions of the solution space. Particular difficulties are encountered in selecting the values of σ, γ and T. Turbulent behaviour (especially bifurcation sequences leading to chaos) has, however, been observed by varying the values of some coefficients of Eq. (6.5) by means of suitable experimental devices (for example, electro-optic modulation of an absorber). In this case, the system (6.5) becomes non-autonomous, but it can be given the form of an autonomous system by increasing the number of the degrees of freedom from two to three.

6.4 OSCILLATING CHEMICAL SYSTEMS

Let us now consider the complex behaviour of those chemical systems exhibiting self-organisation phenomena such as the formation of stationary spatial structures as well as periodic oscillatory states.

Here we will only cover periodic oscillations which, by themselves, offer an extremely rich image of the dynamic behaviour of the system. The experiments to which we will refer regard continuously stirred systems in which the process of forced diffusion prevents the formation of concentration gradients within the solution.

The periodic oscillations, when they correspond to a limit cycle behaviour, may be considered as self-organisation phenomena: in fact, the behaviour of the system, under these particular conditions, is independent of the initial conditions (determined externally) and is self-determined by the intrinsic characteristics of the system itself. It can be seen that the initial conditions will only influence the behaviour of the system during the transient which precedes the establishment of the oscillating state.

The independence of the asymptotic state of the system upon the initial conditions is a general property which clearly characterises self-organising systems and distinguishes them from the comparatively simpler mechanical systems.

It is the remarkable gap between the information content of the input "signals" and the response of the system that clearly reveals the active contribution of the internal degrees of freedom of the system itself. In fact, in the case of chemical systems the output "signals", i.e. the oscillatory behaviour of the concentrations of the intermediates, seem to have little connection with the "input" signals which usually aim at maintaining constant concentrations of reactants and products. Periodically forced chemical oscillators have been described by Schneider (1985).

The study of chemical reactions with oscillating kinetics has drawn increasing interest over the last few decades because it is thought that it will be useful in the interpretation of more complex phenomena of temporal and spatial organisation in biological systems. The great generality of such phenomena can be appreciated if one realises that they affect the behaviour of the same system at different levels: from the macroscopic level as, for instance, circadian rhythms or the spatial organisation of different cellular types (the problem of morphogenesis), down to the cellular and subcellular level as chemical regulatory processes (glycolytic oscillations, to name but one).

The existence of oscillating chemical systems was recorded with certainty in the twenties when, in 1921, the American chemist W.C. Bray described the oscillating temporal behaviour of the concentrations in the decomposition reaction of hydrogen peroxide in water and oxygen in the presence of iodine as catalyst. Bray's observation remained an isolated one for two reasons. First, his system was not easily controllable and reproducible and it was therefore understandable that sceptics considered the oscillations as artefacts due to the presence of various impurities in the solution. Secondly, this scepticism was reinforced by an incorrect interpretation of the second law of thermodynamics.

The occurrence of an oscillation (say the motion of a pendulum) requires that the spontaneous evolution of the process, which is thermodynamically favoured being accompanied by an increase in entropy (the fall of the pendulum), alternates with the opposite process (the rise of the pendulum). During the latter, even in the case where the system does not pass backwards through all the same states, the roles of initial and final states are exchanged. Accordingly, given that entropy is a function of state, its variation will have the same absolute value both in the forward and the backward processes but be of opposite sign. The inversion of the spontaneous process and thus the periodic recovery of the initial state, seems to be in flagrant contrast with the second law of thermodynamics which excludes any reduction of entropy in a closed and isolated system. In the meantime thermodynamics had undergone profound changes: the birth of nonequilibrium thermodynamics extended the methods and concepts of classical thermodynamics to the treatment of open systems subjected to nonequilibrium conditions. Showing that local entropy decreases may occur in open systems far from equilibrium, nonequilibrium thermodynamics contributed towards resolving the apparent incompatibility between the oscillatory regimes of chemical systems and the second law.

The year before Bray's work, however, A.J. Lotka had demonstrated theoretically that a system of two coupled self-catalysing reactions could produce an oscillatory behaviour. Since then the possibility of applying Lotka's scheme to real chemical systems has not arisen, but as the associated kinetic equations are analogous to those used by Volterra for modelling the prey-predator interaction, their study has shown itself to be particularly stimulating in ecological modelling.

A second example of an oscillating chemical reaction was discovered by the Russian chemist B.B. Belousov in 1958. During the process of oxidation of malonic acid by bromate in a sulphuric acid medium, the periodic variation in the concentrations, in this case, could be clearly noted due to the colour changes undergone by the solution. Several years later, the reaction was the subject of a more thorough experimental examination by his compatriot A.M. Zhabotinski who also noticed the onset of a certain spatial organisation. This was demonstrated, if the reaction took place in a test tube or was confined to a thin film, by the presence of an ordered spatial structure manifested by a series of coloured fringes or concentric rings. In the Appendix a simple recipe is given for realizing a modified Belousov-Zhabotinski reaction.

The growing interest in this kind of phenomenon has recently led to a systematic study of the conditions to be met for the oscillatory behaviour to set in. As a necessary condition a nonequilibrium state must be externally maintained. In general, the most directly controllable parameter is the size of the flow of the reactant mixture which is continually added to the system. The flow is inversely proportional to the so-called residence time of the mixture in the reactor which determines the time available for the reaction to take place. Adopting the terminology introduced in Chapter 5, the flow of reactants or, alternatively, the residence time, plays the role of a bifurcation parameter.

By acting upon this flow one can direct the system towards more or less marked states of nonequilibrium. If the flow tends to zero, then the system will be stabilised with equilibrium conditions which are compatible with the reactant concentrations and vice versa if the flow tends to infinity, then the residence time tends to zero, and therefore, the equilibrium of the reaction moves towards the reactants.

Two additional conditions which seem to favour the establishment of an oscillatory behaviour, are the presence of feedback mechanisms (typical in self-catalysing reactions) and the intrinsic bistability of the system (i.e. the presence of two stable stationary states). An analysis of these conditions is of primary importance for the design of chemical oscillators. The three conditions outlined above have been demonstrated to be valid as heuristic instruments as they have allowed the discovery of one new class of chemical oscillating systems known as chlorite oscillators.

The continuing effort towards the isolation of the minimum requisites for a system to show oscillatory behaviour has led to the construction of a number of simplified theoretical models.

On the basis of the reaction scheme proposed by Field, Körös and Noyes for the Belousov-Zhabotinski reaction, successfully compared with the experimental data, a reduced model has been proposed, known as the Oregonator, which includes only three kinetic equations. The Oregonator is capable of accounting for a certain number of properties of the original model, and in particular predicts the establishment of periodic behaviour. The Oregonator has also been used to generate a certain number of more refined models which succeed in describing the birth of chaotic situations. The analysis of the Oregonator is rather tedious and thus it is better to begin our study using more simple and manageable models. Historically, the predecessor of all chemical oscillating systems is that proposed by Lotka in 1920, which consists of a set of three hypothetical reactions:

$$A + X \longrightarrow 2X$$
$$X + Y \longrightarrow 2Y \qquad (6.36)$$
$$Y \longrightarrow P$$

The kinetic equations for the intermediates X and Y are then given by:

$$\frac{dx}{dt} = Ax - xy$$
$$\frac{dy}{dt} = xy - y \qquad (6.37)$$

which are formally analogous to the Volterra equations (5.17). Under nonequilibrium conditions the system allows an oscillating regime corresponding to the possible trajectories about a stationary point which turns out to be a center (see Section 5.3 and Fig. 5.10). These oscillations, however, are extremely sensitive to the initial conditions in the sense that on varying the latter the representative point moves along increasingly different orbits. The first (albeit hypothetical) system of reactions which showed an oscillatory behaviour corresponding to a limit cycle is the so-called Brusselator, which is also the simplest example of a system admitting limit cycle behaviour. In order to show the existence of a limit cycle, the instruments introduced in Chapter 5 will be applied to this system.

The Brusselator is associated with the following sequence of reactions:

$$A \; \longrightarrow X$$
$$B + X \longrightarrow Y + D$$
$$2X + Y \longrightarrow 3X \qquad\qquad (6.38)$$
$$X \; \longrightarrow E$$

Adding Eqs. (6.38) we obtain

$$A + B \; \longrightarrow \; D + E \qquad\qquad (6.39)$$

Therefore, in the complete process, X and Y have the role of intermediates, the only function of which is that of mediating the conversion of the reagents A and B into the products D and E without undergoing any permanent transformation at the end of the reaction cycle. Usually the concentrations of the reagents and the products are kept constant and the course of the reaction in time is followed by monitoring the concentrations of the intermediates.

It is noted that in all the reactions (6.38) the reverse reactions have been ignored. These conditions of irreversibility may be realised by driving the system far from equilibrium, i.e. by imposing $ab \gg de$ (lower case letters denote the concentration of the corresponding compound). For the affinity of the process

$$\mathscr{A} = RT \ln \frac{ab}{de} \qquad\qquad (6.40)$$

we obtain

$$\mathscr{A} \gg 1 \qquad\qquad (6.41)$$

The Brusselator is also known as the trimolecular model precisely because of the character of the third reaction. The principle reason which suggested its insertion into the reaction scheme (6.38) resides in the following theorem: a sequence of reactions involving two intermediates does not allow a limit cycle about an unstable focus if all the reactions are of a monomolecular or a bimolecular type.

The trimolecular reaction also represents the autocatalytic step which, together with the condition of nonequilibrium (guaranteed by Eq. (6.41)), is one of the recurring characteristics of oscillating chemical systems.

One could object to the validity of including in the model such an improbable event as a trimolecular reaction. This objection may be overcome by taking

$$2X + Y \; \rightleftarrows \; 3X \qquad\qquad (6.42)$$

and dividing it into two bimolecular steps

$$X + X \; \rightleftarrows \; XX \qquad\qquad (6.43\text{a})$$

$$XX + Y \longrightarrow \; XX + X \qquad\qquad (6.43\text{b})$$

where XX is the dimerised form of X. Now it can be demonstrated that given suitable restrictions on the rate constants, the kinetic equations relating to the scheme (6.43) will, on the whole, reproduce the kinetics of the reaction (6.42).

Let us now go back to Eqs. (6.38) and, having put all the reaction constants equal to 1 to simplify the calculation, we can write the kinetic equations for the intermediates:

$$dx/dt = a + x^2 y - bx - x \qquad\qquad (6.44\text{a})$$

$$dy/dt = bx - x^2 y \qquad\qquad (6.44\text{b})$$

Let us begin the study of Eqs. (6.44) by looking for the stationary states. Given the condition that the system is in a stationary state

$$dx/dt = dy/dt = 0 \qquad (6.45)$$

we can identify only a single stationary state having the coordinates

$$\begin{aligned} x_0 &= a \\ y_0 &= b/a \end{aligned} \qquad (6.46)$$

whose stability will be studied as a function of the two parameters a and b. It will be assumed, for obvious physical reasons, that $a > 0$ and $b \geqslant 0$.

The Jacobian matrix (Eq. (5.26)) associated with the system of Eqs. (6.43) and (6.44), at the point of equilibrium, is

$$\mathbf{J}\left(a, \frac{b}{a}\right) = \begin{Vmatrix} b - 1 & a^2 \\ -b & -a^2 \end{Vmatrix} \qquad (6.47)$$

and takes the eigenvalues:

$$\lambda_{1,2} = \frac{-(a^2 + 1 - b) \pm \sqrt{(a^2 + 1 - b)^2 - 4a^2}}{2} \qquad (6.48)$$

If

$$a^2 + 1 - b > 0$$

then both the eigenvalues are either real and negative, or complex conjugate with a negative real part. In both these cases the equilibrium point is stable; it turns out to be unstable if

$$a^2 + 1 - b < 0$$

(see Fig. 6.6).

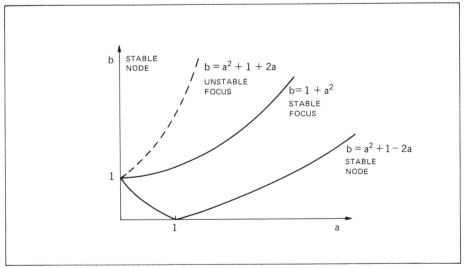

Fig. 6.6 - Stability and type of equilibrium point (Eq. (6.46)) for the system described by Eqs. (6.43) and (6.44) as a function of the parameters a and b.

In the plane spanned by the parameters a and b the parabola $b = 1 + a^2$ divides the positive quadrant into two regions where the equilibrium point is stable or unstable (see Fig. 6.6). Noting that the discriminant in Eq. (6.48), vanishes for $(a^2 + 1 - b)^2 = 4a^2$, one can say that the two parabolas of equation $b = a^2 + 1 \pm 2a$ divide the positive quadrant into regions where the equilibrium point is respectively a node or a focus (see Fig. 6.6).

Corresponding to each value of the parameters a and b, the eigenvalues assume imaginary values on the parabola $b = 1 + a^2$. Therefore, with a fixed value of a, $b^* = 1 + a^2$ is a bifurcation value, as for $b < b^*$ the equilibrium point is stable, for $b > b^*$ the equilibrium point is unstable and for $b = b^*, Re\,(\lambda_{1,2}) = 0$. It can also be noted that

$$\left. \frac{\partial Re(\lambda_{1,2})}{\partial b} \right|_{b=b^*} = 1 > 0$$

and thus it is interesting to discover whether the condition of Eq. (5.53) which, according to the Hopf theorem, assures the birth of a stable limit cycle for $b > b^*$, is met. Towards this end, a coordinate transformation is carried out on Eqs. (6.43) and (6.44), in order to shift the equilibrium point (see Eq. (6.46)) to the origin, followed by the canonical transformation of Eq. (5.43). The system thus obtained, for $b = b^*$, is:

$$\begin{aligned}
\dot{\xi}_1 &= a\xi_2 + 2a^2\xi_1\xi_2 + a(1 - a^2)\xi_1^2 + a^2\xi_1^2\xi_2 - a^4\xi_1^3 \\
\dot{\xi}_2 &= -a\xi_1
\end{aligned} \tag{6.49}$$

For each value of a, Eq. (5.53) then furnishes

$$A = -\frac{a^2}{8}(2 + a^2) < 0$$

Consequently, the Hopf theorem assures the existence of a stable limit cycle for each value of a and for $b > 1 + a^2$, provided that b remains fairly close to $1 + a^2$. It should be remembered, in fact, that the Hopf theorem has local validity.

Numerical investigations confirm Hopf's predictions. In particular, Fig. 6.7 shows the results obtained for $a = 1$, $b = 2$; $a = 1$, $b = 2.2$; $a = 1$, $b = 3$. In the first case, the equilibrium point $(a, b/a)$ is a stable focus; in the second case it can be clearly seen that this point is unstable and that a limit cycle appears; in the third case the limit cycle is still present although its form has changed.

The Brusselator exhibits that particular type of periodic regime which gives rise to the so-called relaxation oscillations and which has not only been found in the simulations on the Oregonator and in other model chemical systems but has also been observed in the Belousov-Zhabotinski reaction. In the typical curve for these oscillations, shown in Fig. 6.8 for the intermediate X, rapid and conspicuous variations can be seen between periods of almost imperceptible variability. This kind of discontinuous oscillatory behaviour is generated whenever the conditions

$$\begin{aligned}
a &= finite \\
b/a &\to \infty
\end{aligned} \tag{6.50}$$

are met, which imply the instability of the stationary state, as can be seen from Fig. 6.6.

It will be useful, for the following analysis, to adopt the new variable $z = x + y$ in order to transform Eqs. (6.43) and (6.44) into

$$\dot{z} = a + y - z = F(y, z) \tag{6.51}$$

$$\epsilon\dot{y} = x - \epsilon x^2 y = (z - y)[1 - \epsilon y(z - y)] = G(y, z)$$

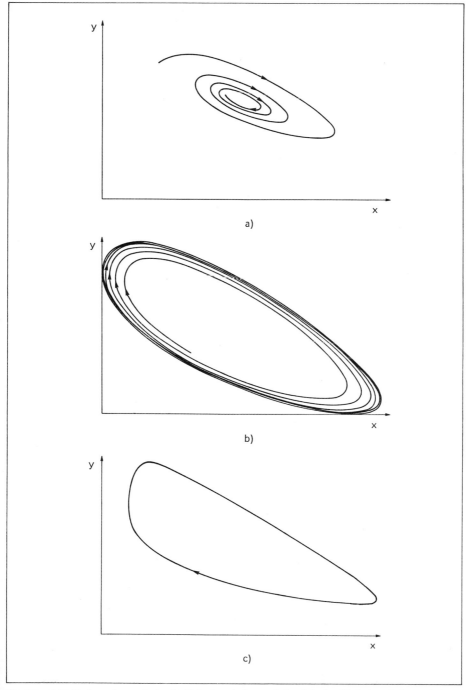

Fig. 6.7 - a) Stable focus for $a = 1$, $b = 2$; b) limit cycle for $a = 1$, $b = 2.2$; c) stable limit cycle for $a = 1$, $b = 3$.

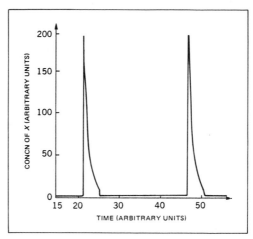

Fig. 6.8 - Relaxation oscillations from the Brusselator monitored on the intermediate X (after Lavenda, Nicolis and Herschkowitz-Kaufman, 1971).

where $\epsilon = 1/b$.

The continuous curve in Fig. 6.9 shows the form that the limit cycle, corresponding to the relaxation oscillations of Fig. 6.8, will take when $b/a \to \infty$. The approximate shape of the limit cycle may be predicted on the basis of Eqs. (6.51) and (6.52), under the conditions expressed by Eq. (6.50). It can be seen that for the condition given by Eq. (6.50) if $G(y, z) \neq 0$ then Eq. (6.52) predicts $\dot{y} \gg 1$ whereas $\dot{y}$ will take finite values when $G(y, z) = 0$. From a geometric point of view, recalling that the trajectories of the phase plane have the equation

$$\frac{dz}{dy} = \epsilon \, \frac{F(y, z)}{G(y, z)} \tag{6.52}$$

it can be seen that these phases, having very fast kinetics for y, correspond to segments (BC and DE in Fig. 6.9) which are almost parallel to the y axis. In fact, $G(y, z) \neq 0$ and $\epsilon \ll 1$ imply that $dz/dy \approx 0$. The phases with slower kinetics occur when $G(y, z) \approx 0$ or in the neighbourhood of the curves

$$z_1 = y$$
$$z_2 = y + b/y$$

Figure 6.9 shows the curve z_2 (whose right and left branches join in the vertex B) and the bisector z_1 which is also asymptotic with z_2. U is the unstable stationary point (Eq. (6.46)), located at the intersection of z_2 with the line r, whose equation is $z = a + y$. The dashed or continuous branches of the various curves represent, respectively, their "repulsive" or "attractive" property, namely the fact that the velocity vector ($\dot{y}, \dot{z}$) diverges from the dashed branches and converges towards the continuous ones. Away from the neighbourhood of the curves z_1 and z_2 (within which $G(y, z) \approx 0$), the y component dominates with respect to the z one (in fact $\dot{y} \gg 1$ and $\dot{z} \approx 0$) so that the velocity of the representative point will be roughly parallel to the y axis (as near the branches BC and DE). This means that starting from a general initial condition A there will be a rapid movement alongside the closest "attractive" branch (the branch EB in Fig. 6.9), following a path (AF) almost parallel to the y axis. While keeping close to EB, the representative point moves towards the vertex B of the parabola since in this region (above the line r) $\dot{z} < 0$.

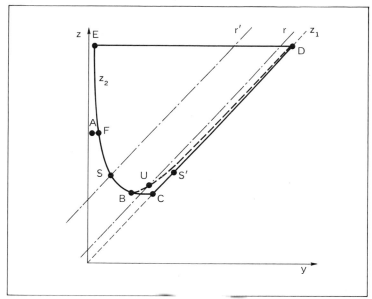

Fig. 6.9 - **Limiting shape of the cycle corresponding to the relaxation oscillations of Fig. 6.8 which are produced when the stationary state is unstable (point U). Whenever the stationary state is stable (point S), fluctuations in y of sufficient intensity may cause the representative point to jump onto the branch CD. In this case, the return to the state S takes place along the branches CD, DE and ES and thus leads to a significant transient variation in the concentration of the intermediates (threshold excitation) (after Lavenda, Nicolis and Herschkowitz-Kaufman, 1971).**

Having passed the vertex (y being greater than zero) and crossing the line r (along BC) the point enters the region in which $\dot{z} > 0$ thus returning upwards while remaining close to the bisector. The return upwards undergoes a change when the trajectory intersects the dashed branch of the parabola so that it enters in the $\dot{y} < 0$ region. (It can be shown, in fact, that the trajectory does not remain indefinitely trapped between the branch of the parabola and the bisector). The point then moves left and rapidly returns towards the branch EB along DE entering into a new descent stage. The correspondence between the oscillations shown in Fig. 6.8 and the approximate shape of the cycle in Fig. 6.9 is as follows:

along DE $z \approx$ constant, y decreases and x increases;

along EB z decreases, $y \approx$ constant and x decreases;

along BC $z \approx$ constant, y increases and x decreases;

along CD $z \approx y$ and $x \approx$ constant.

Thus the sharp peaks correspond to motion along $DEBC$ whereas the stage with very little variation regards motion along CD.

The analysis of the velocity field in the phase plane (y, z) gives results analogous to those above, and such an analysis is useful for explaining the so-called excitability of the system (about the interesting connection between excitability and pattern formation see Winfree and others, 1985, and references therein). The system is said to be excitable when, in response to perturbations of an intensity beyond a characteristic threshold, the return to the stationary state is preceded by a large oscillation of the variables around the stationary values.

Let us return to Fig. 6.9 and consider the stable stationary state S which is again located at the intersection of the straight line r' and the parabola but now lies on the stable branch *EB*. The destabilising action of the perturbation (or of the fluctuation) provokes the transition from the state *S* to the state *S'*, in the neighbourhood of the branch *CD*, so that the return to the state *S* must occur along a path which tends asymptotically towards the trajectory *S'DES*.

The time evolution of the variable *x* during the transient will be similar to one of the oscillations in Fig. 6.8; it must be noted that the initial negative perturbation, corresponding to the transition *SS'*, leads to a positive amplified response (*x* increases along *DC*). The Oregonator also turns out to be excitable and numerical simulations have shown that a perturbation of 6% in the concentration of one of the intermediates causes, in response, transient variations which may even be several orders of magnitude greater than the initial perturbation (Field and Noyes, 1974, and Schneider, 1985).

6.5 ORDER AND CHAOS IN CHEMICAL SYSTEMS

In Section 6.4 the Brusselator was used as a suitable model for carrying out a semi-analytical study of simple oscillatory behaviour which is frequently observed in a variety of chemical and physical systems.

In this Section, the study is extended to oscillatory regimes of greater complexity which are frequently found in real chemico-physical systems. An examination of the phenomenology of the latter will also allow the consideration of chaotic regimes which, in these systems, unlike the case of the laser, can be observed with little difficulty. Finally, the subject of chaos in deterministic systems will provide precious elements for a reconsideration of the counterposition of stochastic and deterministic systems and for classifying new deterministic behaviours as well as other noise-induced effects.

The investigation of chaotic regimes is of current interest since the discovery that even systems of deterministic equations (such as, for example, those cited in Sections 5.7 and 6.3) can give rise to unexpected disordered, chaotic or "turbulent" behaviour, as it is also defined according to a terminology originally used for describing chaotic behaviour in fluid dynamics. Under such conditions of extreme disorder the system may "forget" its own initial conditions over even very short times, a property which has previously been traced back to the presence of fluctuating terms.

It is therefore important to compare stochastic and deterministic chaos in order to draw out the undeniable analogies, but also some fundamental differences. Secondly, it is also useful to discuss the difficulties arising when one faces the problem of detecting deterministic chaos experimentally, starting from experimental data, and finally, to investigate the combined effects of stochastic and deterministic noise.

From this point of view, the chemical system most widely studied experimentally is the Belousov-Zhabotinski reaction. Any comparison with the theoretical predictions is of course problematic due to the fact that the detailed reaction mechanism is still the subject of debate and, consequently, the kinetic equations proposed should not be considered as definitive ones. In spite of the uncertainty affecting the model equations, this task is not hopeless. In fact it should be stressed that the more general characteristics of the various types of transitions to chaos are relatively independent of the particular aspects of the starting model and therefore, in this preliminary stage of classification of chaotic behaviour, it is not necessary to use excessively sophisticated models.

The generality of such behaviour is confirmed by the profound analogies which exist in the sequence of states in such diverse systems as the laser, thermoconvective instability (Bénard cells) and, as already mentioned, certain classes of chemical systems.

Before going into the genesis of various types of chaotic regimes, and of the possible se-

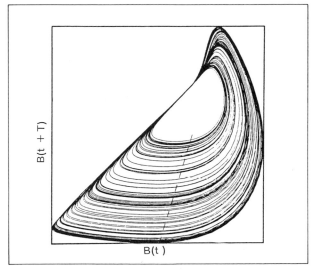

Fig. 6.10 - Two-dimensional phase portrait of a chaotic state of the Belousov-Zhabotinski reaction (after Roux and Swinney, 1981).

quences of chaotic and oscillatory regimes, it is worth examining some additional notions which will prove useful in the characterisation of the various types of attractors, including both periodic (limit cycles) and non-periodic (chaotic) attractors. This will result in the quantification of the disorder associated with chaotic states and will then provide a more rigorous definition of deterministic chaos.

Let us return to the Belousov-Zhabotinski reaction in order to follow in more detail the characterisation of the chaotic states both from a theoretical and from an experimental point of view. Figure 6.10 represents the two-dimensional projection of a three-dimensional phase portrait, reconstructed on the basis of experimental data relating to a chaotic state of the reaction. A discussion of the choice of the variables and a justification of the particular type of representation adopted will be given later. Here we will only note that the preferred variables are the potential values of one of the intermediates (the bromide ion) at three different points in time ($B(t)$, $B(t + T)$ and, perpendicular to the plane of Fig. 6.10, $B(t + 2T)$; attention will be paid, for the moment, to some of the relevant properties of this chaotic attractor.

Following the procedure outlined in Section 5.7, the Poincaré section is constructed along the dashed line in Fig. 6.10. Figure 6.11 shows that the points of the section are practically aligned. It is precisely this laminar character of the section, highly ordered and reproducible, which is one of the indications, although not a rigorous proof, of the fact that the chaotic motion along the attractor is of a deterministic rather than of a stochastic nature. If the "turbulent" behaviour had been caused by intense fluctuations, the alignment of the points on what is, within experimental error, a reproducible curve, would be difficult to explain.

Given a stable periodic orbit, the temporal behaviour of the variables is periodic, apart from deviations from the ideal orbit caused by possible fluctuations. This will be even more the case, if we consider slight perturbations of only the initial conditions. In the periodic case, initial differences are not expected to bring about radical changes in the successive evolution of the system which, following a transient, will re-acquire a periodic behaviour. The reproducibility of the curve in Fig. 6.11 (a) would seem to indicate an analogous course even in the

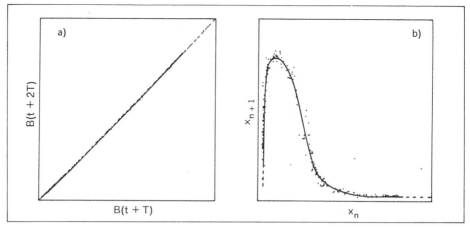

Fig. 6.11 - (a) Poincaré section obtained by dissecting the attractor of Fig. 6.10 along the dashed line with a plane perpendicular to the plane of the page. (b) Poincaré return map, constructed by taking as abscissa the coordinate $B(t + T)$ of each point of the section in Fig. 6.11 (a) and, as ordinate, the coordinate $B(t + 2T)$ of the immediately succeeding point in time (after Roux and Swinney, 1981).

chaotic regime. But this is not the case because, although the curve is reproducible, it is impossible to regenerate exactly the same series of points along it.

It is rather the dispersion of the experimental points around the solid line in Fig. 6.11 (b) which shows the effect of the fluctuations. The latter are caused, for example, by the presence of impurities or by changes in stirring speed, which for a very brief period altered the composition of the flow of reactants which, in theory, should remain rigorously constant.

A partial confirmation of this is given by following the evolution of the system after an artificially induced perturbation. Figure 6.12 shows the wanderings of the representative point along the section and on the Poincaré map, the system having been perturbed after the fourth iteration (for a review of similar experiments see Schneider, 1985).

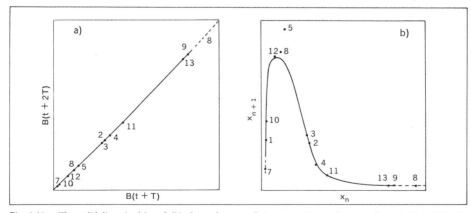

Fig. 6.12 - The solid lines in (a) and (b) show the same Poincaré section and map as those in Fig. 6.11. The numbers indicate the position of the successive iterates, estimated from an experimental sequence of data; the spontaneous evolution of the system has been artificially perturbed at point 4 (after Roux and Swinney, 1981).

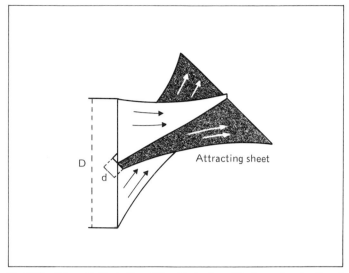

Fig. 6.13 - Schematic illustration of the effects of amplification and contraction of the distances observed upon a strange attractor and in its immediate neighbourhood.

Points 5 and 8 effectively move away from the map trace (continuous curve in Fig. 6.12(b)). The fact that the anomaly is not visible in the section of Fig. 6.12(a) shows that the approximately two-dimensional manifold defined by the attractor is strongly attractive. This means that perturbations which take the system outside of the "surface" are promptly neutralised so that, even after a single iteration, the representative point has returned to the original section (at least within the limits of the resolution of Fig. 6.12). Only the components of the perturbations which lie upon the "surface" of the attractor survive and these are even susceptible to amplification. As we shall see below in more detail, a typical feature of the motion of the representative point along chaotic attractors is in fact the extreme sensitivity to initial conditions.

The effect of the perturbation would be detectable only if we were able to place the system, with infinite precision, in exactly the same initial condition as in the experiment in Fig. 6.12. By comparing the positions of the various intersections of the perturbated and the unperturbed trajectories, it would then be possible to verify whether the two series would coincide up to the fourth iteration before clearly diverging.

Although round-off errors prevent us from starting the system from the same initial condition, the map in Fig. 6.12 (b) can detect the presence of the fluctuation. In fact, the map contains information about the exact succession in time of the iterates; therefore the interference of the fluctuation is revealed by an anomalous shift in the curve position, in the iterations immediately following the perturbation.

Figure 6.13 shows schematically the "attractive power" of the attractor. The rapid convergence of points initially placed at a distance $D/2$ from the attractor towards the "surface" of the attractor (shaded area), indicates the decay of those fluctuations which continually tend to take the representative point away from the attractor. On the contrary, the distance d, although very small, between two points which are initially on the attractor, is susceptible to amplification due to the marked local divergence of the orbits. It is this mechanism which causes the failure of attempts at reproducing the evolution of the system. In fact, the inevitable approximation, which is inherent in every computer simulation, is amplified so that the subsequent behaviour of the

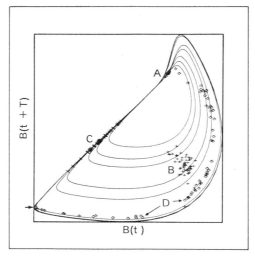

Fig. 6.14 - Evolution of a cloud of points having almost
the same initial conditions. The initial volume of
the phase space filled by the cloud is practically
one point, situated near the arrow at the bottom left of the chaotic attractor. The different
snapshots (or "strobes") ($A(\triangle)$, $B(+)$, $C(\backslash)$ and
$D(\diamond)$) show the position and spreading of the
cloud at succeeding points in time (after Roux
and Swinney, 1981).

system becomes unpredictable. Given similar initial conditions it would then be fallacious to
deduce that the system's successive evolution will be similar.

This characteristic effect of strange attractors is responsible for the loss of memory by
the system of its initial conditions which distinguishes chaotic regimes. Fig. 6.14 provides an
example of this phenomenon obtained from a simulation using a four-dimensional model of the
Belousov-Zhabotinski reaction. During the course of the simulation the evolution of a cloud of
points, initially enclosed within a small volume (whose position is indicated by the pointer), was
obtained. The case in which the initial conditions of the various points differ by quantities
which are well below the numerical or experimental sensitivity, was simulated. The points appear
to be coincident and the small volume is indistinguishable from a point in the plane of Fig. 6.14.
Nonetheless, the above dispersive effect comes to the fore, eventually spreading the points over
a wide region of the attractor.

Fig. 6.15 gives an idea of the rapidity of the phenomenon. A map analogous to that in
Fig. 6.11, obtained from experimental data, was used to calculate the dispersion of the values of
the bromide ion potential as a function of the number of iterations, starting from a set of ten
values lying between 159.5 and 160.4 (on the scale of the diagram this interval corresponds to
a single point). After only five iterations there is a significant dispersion of the points which
progressively tend to spread over a much larger interval.

At this point it should be stressed that the spreading over the attractor of points having
initial conditions which are even very similar to each other, the so-called "stretching" effect,
has a natural limit due to the finite extension of the attractor. The fact that almost all the
small starting volumes have transforms which, after only a few iterations, diffuse over the attractor and therefore tend to cover the same numerical intervals, together with the fact that these

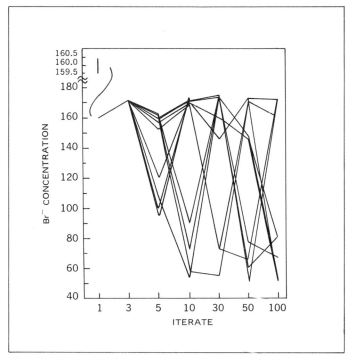

Fig. 6.15 - Divergence of the trajectories as a function of the number of iterations visualized on a Poincaré map for a chaotic state of the Belousov-Zhabotinski reaction (after Hudson and Mankin, 1981).

transforms must, in any case, be distinct in order to safeguard the uniqueness of the solution, implies that the attractor must take on a "folded" structure. This is the only way in which it is possible to have overlapping "sheets" which are, however, distinct along the direction of their "thickness".

On cutting the attractor in Fig. 6.10 at successive positions, a series of sections is obtained which is schematically illustrated in Fig. 6.16. It is clear that it is the folding of the attractor which is responsible for the marked dispersion of the iterates along the section in Fig. 6.12 (a), particularly accentuated for iterates 9 to 13. The shape of a trajectory starting from a point

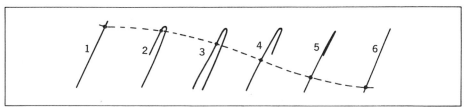

Fig. 6.16 - A sequence of Poincaré sections obtained at successive positions in order to show the "folding" of the chaotic attractor in Fig. 6.10; the dashed line represents a stretch of a possible trajectory. Section 6 coincides with section 1 and is given in order to underline the drift which the representative point may undergo during the course of a single iteration, due to the folding (after Roux and Swinney, 1981).

situated at the top of section 1 (say point 9 in Fig. 6.12(a)) has been sketched, joining the positions of the representative point on the successive sections. It can be seen that, due to the folding, the trajectory crosses section 1, after a single iteration, in a diametrically opposed position (see point 10 in Fig. 6.12(a)).

This property of the attractor can be correlated with the characteristics of the associated map, as it may be stated that a necessary condition for the occurrence of the stretching property is the noninvertibility of the map (i.e. the presence of at least one maximum in the plot of the function).

A particularly clear example is given by the map

$$x_{n+1} = f(x_n) = a\,(1 - 2\,|x_n - 1/2\,|)\,, \qquad 0 < a \leqslant 1 \qquad (6.53)$$

Fig. 6.17(a) shows the plot of the map for $a = 1$ and Fig. 6.17(b) gives the trends of the m-th iterates (obtained by applying m times the map (6.53) to x_n). It can be seen that with an increasing number of iterations, progressively smaller intervals (of amplitude 2^{-m}) on the x_n axis are mapped over the same interval [0,1]. As a consequence, after a sufficient number of iterations, arbitrarily close points will have transforms which tend to disperse over the same finite interval. We can thus conclude that the stretching property gives rise to the extreme sensitivity of strange attractors to initial conditions. From Fig. 6.17(b) it is clear that from initial conditions which differ by arbitrarily small quantities, trajectories may emerge which are so divergent that they are to be found in completely different regions of the attractor.

On the other hand, the system may be forced to explore the neighbourhood of the same point of the attractor starting from very different initial conditions. In Fig. 6.17, it is shown how an increasing number of points initially enclosed within the same starting interval possesses the same m-th iterate. Each noninvertible map which is used for modelling chaotic states of deterministic systems would thus seem to contradict the theorem of uniqueness of the solution as it would predict, after m iterations, the intersection of orbits emerging from different initial conditions. This is, however, only an apparent contradiction since it should be recalled that the one-dimensional map is a reduced description of an attractor which is actually immersed in a space of greater dimensionality. In this space, the iterates which appear to be coincident on the one-dimensional map correspond to points which may be distinguished by means of the remaining coordinates.

Therefore, in practice, it would seem that systems in chaotic regimes were capable of forgetting their own initial conditions. The break between the past and the future thus occurring in the chaotic state recalls the situation of stochastic systems where fluctuations can cause the

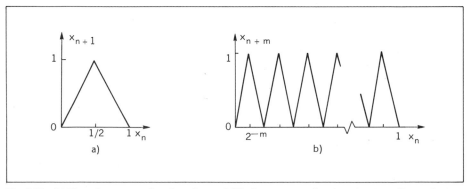

Fig. 6.17 - (a) Plot of x_{n+1} as a function of x_n and (b) of x_{n+m} as a function of x_n according to the map (6.53).

system to evolve towards states which are independent of the initial conditions (after a sufficient time lapse). Although the effects on the state variables is analogous in both cases, it must be stressed that the physical origin of the disordering mechanism is different. The deterministic case originates from the time evolution equations for the degrees of freedom of the system, whereas fluctuations regard the interaction of the system with external degrees of freedom.

After having outlined the qualitative characteristics of the spreading of the points over the attractor and having defined the connections with some properties of the corresponding map, it is worth examining the quantities which allow a rigorous quantification of these properties. The definition of parameters of this type will prove to be very useful for providing a more formal characterisation of chaotic attractors and also for discriminating unequivocally between periodic and chaotic attractors.

In the following we will not discuss the more abstract characterisation of chaotic attractors in terms of fractal dimensions, and will not examine the possible connection with information theory concepts. Detailed accounts of these topics may be found, for example, in the papers by Benettin and others (1976), Blacher and Perdang (1981), Eckman and Ruelle (1985), Farmer and others (1983), Grassberger and Procaccia (1983.b), Show (1981), Young (1983) (see Bibliography). We will rather privilege those approaches which are more accessible to physical intuition.

In the light of the above-mentioned analogy with stochastic systems it is quite natural to make use of the autocorrelation function of the signal as a proper indicator of the "degree" of chaos of the actual state of the system. If an experimental time series is available for the variable $x(t)$, the function can be defined as

$$C(\tau) = \langle [x(t) - \langle x \rangle] [x(t + \tau) - \langle x \rangle] \rangle / \langle [x(0) - \langle x \rangle]^2 \rangle \qquad (6.54)$$

whereas when dealing with a one-dimensional map, the function will be defined as

$$C(m) = \lim_{N \to \infty} \frac{1}{N} \sum_{n=1}^{N} (x_n - \langle x \rangle)(x_{n+m} - \langle x \rangle) \qquad (6.55)$$

where $\langle x \rangle$ is the average value of the N recorded values or iterates

$$\langle x \rangle = \lim_{N \to \infty} \frac{1}{N} \sum_{n=1}^{N} x_n \qquad (6.56)$$

As can be seen from Figs. 6.18 and 6.19 the average autocorrelation function exhibits an oscillatory behaviour at periodic states, while it rapidly decays to zero in the case of chaotic states.

In fact, in a chaotic regime the deviations from $\langle x \rangle$ are, on average, correlated only for sufficiently small values of m or τ (as for points 2,3 and 4 in Fig. 6.12(a)) while on increasing m(or τ) the two factors of each term of the summations in Eqs. (6.54) and (6.55) tend to give progressively more uncorrelated deviations which make the sum converge to zero.

Let us now turn to the so-called Lyapunov numbers and Lyapunov exponents which are commonly used to discriminate among periodic and chaotic states. Hereafter we will only sketch the procedure involved in the mathematical definition of the Lyapunov numbers. Technical details may be found in the papers by Benettin and others (1976, 1980) (see Bibliography).

The Lyapunov exponents give the average exponential rates of divergence or convergence of contiguous orbits. Referring back to Fig. 6.13, these exponents determine the average rate of exponential growth or decrease with time of the initial distances d and D, respectively, between two pairs of points moving on the attractor or in its neighbourhood.

The procedure for constructing the Lyapunov numbers will be described in more detail by considering a three-dimensional space which will be sufficient for all the examples which follow. First of all, a small spherical volume, of infinitesimal radius d, in phase space will be

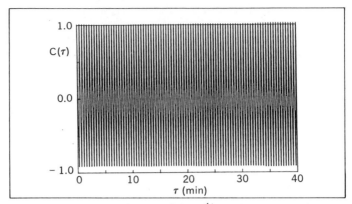

Fig. 6.18 - **Autocorrelation function of the Ce^{4+} ion absorption at 340 nm in a periodic state of the Belousov-Zhabotinski reaction (after Vidal and others, 1980).**

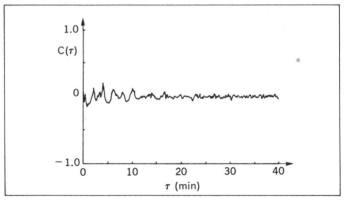

Fig. 6.19 - **Behaviour of the function (6.54) for a chaotic state of the Belousov-Zhabotinski reaction (after Vidal and others, 1980).**

defined to represent, as in the simulations referred to in Figs. 6.14 and 6.15, a set of infinitely close initial conditions and the time evolution of the points within it will be followed. Over sufficiently brief time intervals the sphere will generally be deformed into an ellipsoid whose three principal axes will indicate the directions along which the infinitesimal sphere has expanded or contracted. Let us now iterate the same operation choosing, however, different initial positions of the small sphere along the attractor, and recording each time the ratio between the length p_i (i = 1, 2, 3) of each principal axis and d. The average values of the logarithms of p_i/d thus obtained will provide the three Lyapunov numbers n_i which will then give the average coefficients of expansion or contraction of the sphere radius on the attractor. The Lyapunov exponents λ_i are defined as the natural logarithms of the Lyapunov numbers. The growth of the lengths in phase space and thus the divergence of the trajectories will then be associated with Lyapunov numbers larger than one or with positive Lyapunov exponents. It must be pointed out that the principal axes of each infinitesimal ellipsoid are ordered and named according to their magnitude: for example p_3 will always designate the largest one and p_1 the smallest one (that is $p_3 \geq p_2 \geq p_1$). Accordingly, the same order relation will hold among the Lyapunov exponents (or numbers): $\lambda_3 \gtreqless \lambda_2 \gtreqless \lambda_1$ ($n_3 \gtreqless n_2 \gtreqless n_1$).

The spectrum of the Lyapunov exponents allows one to quantify the "degree" of order or of disorder in the trajectories of the system and turns out to be one of the most adequate and reliable instruments for an unambiguous characterisation of and distinction between the various types of attractors.

The simplest type of attractor, the fixed point, will have a spectrum consisting of three negative exponents which reflect the convergence towards that point of all the orbits emerging from initial conditions located within its basin of attraction.

An attractor of the limit cycle type will have two negative exponents which confirm its attractive power in the direction orthogonal to the cycle. The third exponent will be zero and the direction associated with it will be tangential to the cycle itself. In fact, the periodic nature of the trajectory means that at the end of each period any pair of representative points (however chosen on the cycle) will return to their starting positions. Thus, in the direction of the cycle, there will be, on average, no separation or contraction of contiguous initial conditions and the distance between the representative points should, on average, be preserved.

For a discussion of the spectrum associated with a chaotic attractor, reference will be made to two approximately two-dimensional attractors, having the same type of spectrum (with regard to the sign of the exponents): the Lorenz attractor (mentioned in Sections 5.7 and 6.3 regarding the chaotic behaviour of the laser) and the Rössler attractor (which is examined in more detail in Section 6.6).

It is naturally expected that one of the exponents is negative, which will be the one describing the exponential convergence of the orbits towards the attractor starting from initial conditions outside of the attractor's "surface". On the other hand, the phenomenon of the spreading of the trajectories belonging to the "surface", of which we have so far given only a qualitative description, will be associated with at least one positive exponent.

In the Lorenz and Rössler attractors the third exponent is zero but in chaotic attractors of a different structure it is positive. It should be noted that the direction of the three principal axes of the infinitesimal ellipsoid normally vary with the position on the attractor. This does not exclude a minimum of regularity in the disposition of the three axes of the ellipsoid: in the Lorenz attractor the direction of expansion is mainly parallel to the direction of the flow (i.e. the direction identified locally by the trajectories) whereas in the Rössler attractor it tends to be orthogonal to the flow. Thus a strange attractor is characterised by the presence of at least one positive Lyapunov exponent, so that an analysis of the sign of the largest exponent is sufficient for deciding whether the attractor in question is of a chaotic or of a periodic type.

This exponent can be evaluated by starting from the map, associated with a section of the attractor, according to the formula

$$\lambda = \lim_{N \to \infty} \frac{1}{N} \sum_{i=1}^{N} \log_2 \left| \frac{df}{dx}(x_i) \right| \tag{6.57}$$

where df/dx is the derivative of the map at the i-th iteration. The summation in Eq. (6.57) underlines the global rather than the local character of λ, which is a parameter related to a property of the whole attractor. The positive contributions to the summation come from those points on the map where the derivative has an absolute value greater than one. This is not unreasonable given that it is in exactly those regions of the map where the stretching phenomenon occurs. The diagram in Fig. 6.17(b), relating to the map (6.53), clearly shows that for two very close points on the x_n axis the corresponding points on the x_{n+m} axis may be well far apart due precisely to the fact that $df/dx \gg 1$.

In the particular case of the map (6.53) λ is certain to be positive given that $df/dx > 1$ at all points. Naturally for other maps, such as that in Fig. 6.12 for example, the condition $df/dx >$

$>$ 1 will not be uniformly valid over the whole interval so that in the summation (6.57) negative values will also appear which will reduce the resulting value of λ.

It should be remembered that a significant difficulty arises when Eq. (6.57) is applied to maps obtained from experimental data. The value of λ calculated in this manner may be, in fact, quite sensitive to the fitting parameters used for finding an analytical representation for the curve of the Poincaré map. Recently, however, efficient algorithms have been devised which allow reliable estimates of the non-negative exponents or the leading Lyapunov exponent starting from the experimental time series (see for instance Wolf and others (1985), Wright (1984)). Finally we shall mention in passing the possibility of characterising chaotic attractors by means of the so-called correlation exponent; this approach seems to offer considerable computational advantages particularly when dealing with experimental data stemming from high dimensional systems. For a detailed discussion, the reader is referred to the papers by Grassberger and Procaccia (1983a, 1983c) and Ben-Mizrachi (1984) mentioned in the Bibliography.

From what has been said, it is clear that a map of the type in Eq. (6.53) produces a mixing of trajectories such that after only a few iterations in the neighbourhood of each state, orbits coming from the most disparate regions of the attractor are to be found and this is the mechanism responsible for the loss of information about the initial state which occurs typically in chaotic regimes. The rapidity with which such information is lost is correlated with the largest Lyapunov exponent; the information will persist more or less over time according to the size of the exponent. It has been suggested that this grading of the mixing capacity may explain the different power spectrum structures observed in some systems.

We recall that the power spectrum is the Fourier transform of the autocorrelation function of the signal which, in our case, represents the temporal behaviour of the variable chosen for constructing the contracted phase space. The power spectrum is another possible indicator of the rise of a chaotic regime. In fact, while a periodic or quasi-periodic regime is characterised by a spectrum consisting of a series of peaks, chaos is associated with a spectrum with a broadband component indicating the presence not only of well-defined frequencies but also of all the intermediate frequencies. As an example we may recall that the white noise stochastic term possesses a power spectrum which is constant whereas the power spectrum of the velocity of a Brownian particle (see the end of Section 1.5), takes the form of a Lorentzian function (see Fig. 1.3). It has, however, been noted that spectra of a mixed nature exist, in which the sharp peaks survive alongside the broadband noise structure, which reflects, in the evolution of the variable, the presence of a periodic or semi-periodic component superimposed on the chaotic one.

The Lorenz (see Fig. 5.19) and Rössler (see Fig. 6.36) attractors have spectra which show a different degree of order or, according to current terminology, of phase coherence, as shown in Fig. 6.20. The coexistence, seemingly paradoxical, of order and chaos has been observed, for example, in fluid dynamics experiments in which local turbulent motion, responsible for the broad background of the spectrum, is included in an ordered pattern which develops over a larger spatial scale.

The possible phase coherence of an attractor can be observed by an experiment carried out on these two attractors, similar to that shown in Fig. 6.14, in order to compare the rapidity and the entity of the spreading undergone by a cloud of points having very similar initial conditions. The simulation carried out on the Lorenz attractor shows that the set of points takes on a thread-like conformation, of increasing length, along the direction of the flow until it spreads over the whole attractor so that, after a relatively brief period, it loses all information relating to the initial location of the cloud of points. Also, the final distribution of the points is practically independent of the initial position of the cloud on the attractor (see Fig. 6.21).

The evolution of the set of points on the Rössler attractor, on the other hand, occurs in a very different manner. Fig. 6.22 shows eleven snapshots (only two in Fig. 6.22(d)) taken at succeeding points in time. The expansion of the set, which takes on a thread-like aspect along the direction of the positive Lyapunov exponent (approximately orthogonal to the flow), is clearly visualised. Starting from Fig. 6.22(b), the thread begins to wind around itself, which is a sign of the continual transverse spreading, until it reaches a more dilated configuration (Fig. 6.22(d)) which, however, does not extend over the whole attractor. It is this spatial confinement, lasting for a fairly long period of time, which reveals a less effective chaos-producing mechanism than that of the Lorenz attractor and which is responsible for the sharp peaks remaining in the spectrum shown in Fig. 6.20(b).

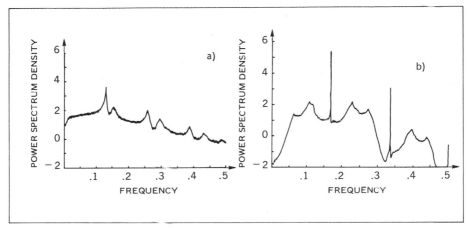

Fig. 6.20 - **Power spectrum density for (a) the Lorenz and (b) the Rössler attractors (after Farmer and others, 1980).**

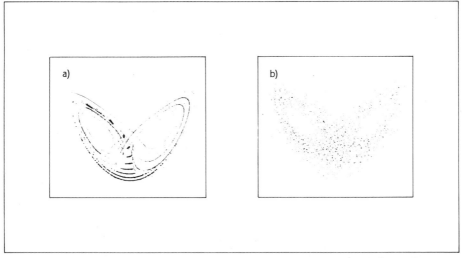

Fig. 6.21 - (a) Ten strobes illustrating the spreading on the Lorenz attractor of a cloud of points having very close initial conditions such that the ensemble is initially indistinguishable from a single point ; (b) "asymptotic" distribution of the same set of points. The Lorenz attractor is lightly outlined in the background (see the full structure of the attractor in Fig. 6.5 (b)) (after Farmer and others, 1980).

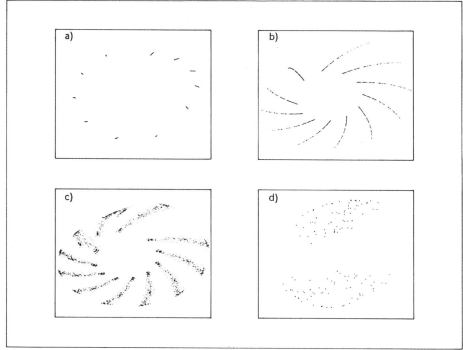

Fig. 6.22 - Successive stages in the mixing process on the Rössler attractor. (a), (b) and (c) each represent eleven (only two in (d)) images which correlate the spreading with the motion of the cloud of points along the attractor (after Farmer and others, 1980).

Let us now return to the problem mentioned in relation to Fig. 6.1 and which is usually met when examining physico-chemical systems whose dynamic equations are unknown or only approximately known. The problem is thus the identification of possible chaotic states starting from experimental data in the form of time series for the various observables.

In the real case of the Belousov-Zhabotinski reaction, the analysis of the system behaviour in phase space meets with serious experimental difficulties due to the large number of variables to be controlled. It would, in fact, be necessary to follow the concentrations of about 30 different chemical components in the reaction. On this point, there are some extremely useful recent results, whose mathematical basis we will not go into here, which fix certain equivalences (e.g. invariance in the spectrum of the Lyapunov exponents) of the original phase space with a space of adequate dimensions constructed on the basis of a single variable.

More precisely, this space is generated by the vectors $B(t), B(t + T), \ldots, B(t + (m - 1)T)$, where $B(t)$ is one of the variables of the system, T is an arbitrary time delay, while m is related to the dimensionality of the original phase space N by the inequality $m > 2N + 1$. Obviously, a space of this type may be easily constructed starting from time series of a single observable.

Although the number of time series to be recorded has been reduced to only one, the remaining difficulty is the examination of a phase space which, for the Belousov-Zhabotinski reaction, would be about 61-dimensional. In many cases, however, it is possible to obtain the fundamental information from a subspace of considerably lower dimensionality. A study of the reliability of this procedure has been carried out on the (three-dimensional) Rössler system, in which comparisons were made between the representation of the original space (see Eqs. (6.59)), and two alternative representations, $(y, \dot{y}, \ddot{y})$ and $(x, \dot{x})$, obtained from the former in the limit $T \to 0$.

In such a system it was found that the largest Lyapunov exponent was extremely reproducible in the three different spaces and that the reliability of the contracted representations is essentially related to the simplicity of the topological characteristics of the attractor. This procedure was used in the study of the chaotic states of the Belousov-Zhabotinski reaction The vectors of the reduced space were obtained from the time series of a variable, assuming the dimension of the space to be that above which there is no appearance of essentially new features in the portrait of the attractor. The two-dimensional representation of the chaotic state of the reaction, discussed in Fig. 6.10, was constructed using this procedure following an accurate examination of the different portraits of the attractor (seen from different angles) which confirmed the "practically" two-dimensional nature of the attractor and thus justified the use of a reduced three-dimensional space $B(t)$, $B(t + T)$, $B(t + 2T)$.

Naturally, only an examination of the power spectrum (see Fig. 6.23) as well as calculation of the largest Lyapunov exponent, estimated on the Poincaré map already shown in Fig. 6.11, allow one to reach the conclusion that we are dealing with a chaotic attractor and not a periodic orbit with a very long period.

The experimental confirmation of the existence of chaotic states for the Belousov-Zhabotinski reaction could provide a means for discriminating between the different theoretical models proposed for describing the system. Modified versions of the Oregonator, containing additional terms for simulating the flows in a stirred flow reactor, do not seem to allow chaotic states. Their existence is however, predicted on the basis of more refined versions of the Oregonator itself.

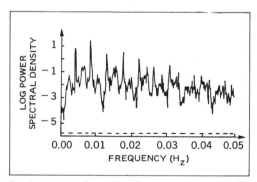

Fig. 6.23 - Logarithmic plot of the power spectral density obtained from a time series of the bromide ion potential for the chaotic state of the Belousov-Zhabotinski reaction corresponding to the chaotic trajectory shown in Fig. 6.10 (after Turner and others, 1981).

Experiment may provide useful indications not only regarding the adequacy of certain models but may also shed light upon a certain number of conjectures regarding the applicability of different formal instruments in the study of these dynamical systems.

In particular, we refer to the adoption of techniques for the reduction of variables such as the use of Poincaré sections and the construction of maps. It is an open question whether or not these simplified representations do indeed preserve all the relevant information about the dynamic behaviour of the system.

In order to answer questions of this type, experimental investigations have been devised not only to verify the existence and the reproducibility of chaotic and periodic states but also to determine their exact sequence in the space of parameters. These checks were also carried out on the Belousov-Zhabotinski system with the intention of comparing the experimental sequence of states with the predictions of the theory of one-dimensional maps and of the kinetic equations which describe the hypothetical reaction mechanism.

One-dimensional maps can be directly obtained from experimental data, as shown in Fig. 6.11, but do not lack attempts to construct abstract mathematical models (i.e. not based upon speculations about the possible chemical mechanism) in terms of *ad hoc* one-dimensional maps which succeed in describing the sequence of states effectively observed in the system .

Studies of this type, focussing on the qualitative predictions of various representations of the same system, are justified by the fact that in this stage a satisfactory qualitative description of the overall behaviour of the system is still lacking, and thus it would be premature to consider the quantitative details of a specific model.

The validity of these attempts may be appreciated by examining Fig. 6.24 where a comparison is made between an experimental sequence of periodic (P_i) and chaotic (C_i) states and the theoretical sequence of states predicted by a modified version of the reversible Oregonator. There is quite a remarkable qualitative agreement, while quantitatively, especially regarding the width of the various windows, the discrepancies could presumably be reduced by a suitable choice of parameters in the model equations.

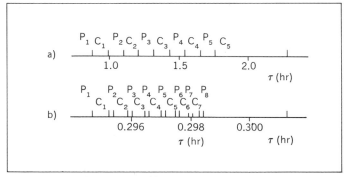

Fig. 6.24 - Comparison between (a) an experimental sequence of chaotic and periodic states of the Belousov-Zhabotinski reaction and (b) the predictions of a four-dimensional reversible Oregonator. The residence time (abscissa) is defined as the ratio between the reactor volume and the feed flow (after Turner and others, 1981).

Under certain conditions the chaotic states (C_i) give rise to behaviours resembling those of the contiguous periodic states $(P_{i-1}$ and $P_{i+1})$; namely, the system executes a random number of P_{i-1} type oscillations followed by a random number of P_{i+1} type oscillations. It should be noted that this type of chaotic behaviour has been attributed to a jumping mechanism between the basins of attraction of two limit cycles very close in terms of the bifurcation parameter. The transition would then be caused by fluctuations and thus the resulting chaos (flip-flop chaos) would be of the stochastic and not deterministic type. So far we have only provided a possible explanation for flip-flop chaos, whereas other types of transitions can be analysed by referring to the properties of unimodal one-dimensional maps or periodic one-dimensional maps.

The best known example of a route to chaos is the transition by period doubling (see Section 5.7) whose characteristics have been studied using unimodal one-dimensional maps. The use of one-dimensional maps as interpretative instruments is justified by the fact that a certain number of states in the period doubling sequence have effectively been observed in the Belousov-Zhabotinski reaction and in a wide range of different systems (e.g. Rayleigh-Bénard convection, nonlinear electrical oscillators) (see Behringer, 1985, Cvitanovic, 1984, Manneville and Pomeau, 1980).

The values of the bifurcation parameter where the various period doublings occur converge, as is known, towards an accumulation point τ_∞, beyond which chaotic regimes and windows of periodic behaviour follow each other according to a characteristic order.

This sequence, known as a U-sequence, consists of a succession of periodic states of period nT ($n = 2, 3, \ldots$), a multiple of the fundamental period T, and of the relative period doubling sequences (with states of period $2^K \, nT$, $K = 1, 2, \ldots$). In the series of states which mark the transition from the P_1 state to the chaotic state C_1 in the sequence given in Fig. 6.24, the first four cycles have been detected, having periods T, $2T$, $2^2 T$ and $2^3 T$ where $T = 115 \, s$. Also, each chaotic window, like that corresponding to the C_1 state, contains the U- sequence whose states of period $6T$, $5T$, $3T$ and $2 \cdot 3T$, have also been observed.

A second route to chaos has been invoked for the interpretation of series of experimental data from systems showing so-called intermittent behaviour. This regime, also observed in the Belousov-Zhabotinski reaction (see Figs. 6.25 and 6.26), includes a "laminar phase" characterised by small oscillations separated by more or less frequent bursts of much larger amplitude.

A plausible and unified description of the transitions described above has been proposed in terms of tangent bifurcations in periodic one-dimensional maps of the type

$$x_{n+1} = ax_n + bP(x_n - x_0) \tag{6.58}$$

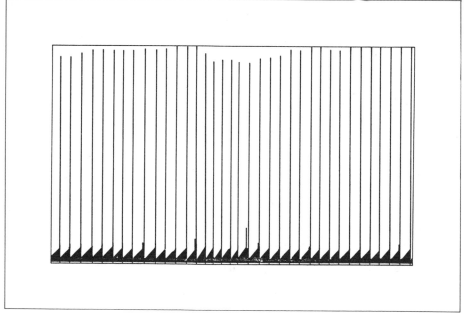

Fig. 6.25 - **Time record of the concentration of one of the intermediates of the Belousov-Zhabotinski reaction** (e.g. Ce^{4+} or the bromide ion) **exemplifying the first type of intermittent regime. (after Roux, 1983).**

where $P(x)$ is a periodic function and the parameters $a(\tau)$ and $b(\tau)$ are monotone decreasing functions of the bifurcation parameter τ (x_0 is a constant parameter).

A simplified scheme of a tangent bifurcation is shown in Fig. 6.27. According to the above hypotheses let us assume that on decreasing the bifurcation parameter τ the curve is shifted upwards. When $\tau > \tau_T$ there will be two separate fixed points, the lower stable and the other unstable, which for $\tau = \tau_T$ merge at the point of the curve tangential to the bisector and finally disappear for $\tau < \tau_T$.

For $\tau < \tau_T$ a channel is opened between the curve and the bisector in which the system is trapped for a certain number of consecutive iterations (as shown schematically in Fig. 6.28). This series of relatively regular oscillations, which corresponds to the "laminar" phase, terminates with a much larger oscillation when the iterate passes the maximum of the curve and, because of the highly negative slope of the descending branch, finds itself below the bisector from where another series of "laminar" oscillations begins (see Fig. 6.28).

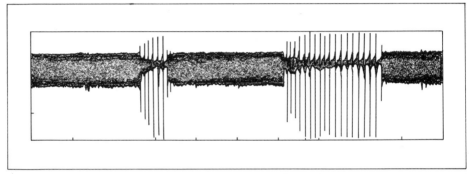

Fig. 6.26 - **Second type of intermittent regime in the Belousov-Zhabotinski reaction, describable in terms of tangent bifurcation (after Roux, 1983).**

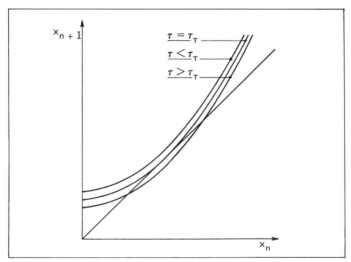

Fig. 6.27 - **Scheme for tangent bifurcation caused by the upward shift of the curve on decreasing the bifurcation parameter τ.**

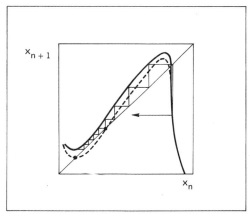

Fig. 6.28 - **Sequence of iterations corresponding to the "laminar" phase typical of the transition to chaos through intermittence (after Roux, 1983).**

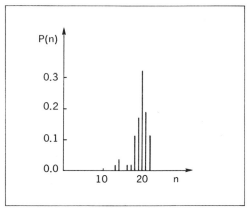

Fig. 6.29 - **Probability function of the number of oscillations in the "laminar" stage (after Pomeau and others, 1981).**

The number of oscillations in the "laminar" stage is not at all constant, even though there is an apparent regularity shown by the signal in Fig. 6.25, but varies according to a probability function, characteristic of this intermittent regime, like the one shown in Fig. 6.29.

Intermittent regimes, similar to that shown in Fig. 6.25 for the Belousov-Zhabotinski system, have also been observed in the Lorenz model and in Rayleigh-Bénard experiments (Pomeau and Manneville, 1980, Manneville and Pomeau, 1980). In this case, the chaotic nature of the regime was confirmed by checking that the largest Lyapunov exponent passes from negative to positive values at the tangent bifurcation.

Experimental studies carried out on the Belousov-Zhabotinski system have also demonstrated the existence of a limit cycle, before the first intermittent regime, corresponding to the fixed point A in Fig. 6.28, which then disappears after the tangent bifurcation.

Under different experimental conditions the destabilisation of the initial periodic state A

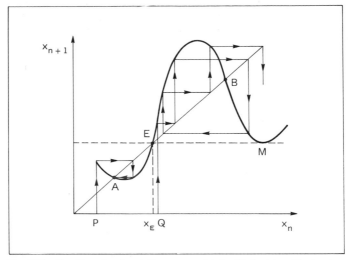

Fig. 6.30 - **One-dimensional periodic map associated with the intermittent regime of Fig. 6.26. The iterations constructed on the curve show that the representative point converges towards the stable fixed point A if the starting point P has an abscissa less than x_E (and thus lies in the basin of attraction of A), or is confined within the basin of the chaotic attractor when the abscissa of the starting point Q is larger than x_E (after Roux, 1983).**

gives rise to an intermittent pattern of the type shown in Fig. 6.26, consisting of "laminar" stages of longer duration and a practically imperceivable amplitude modulation. A qualitative explanation of the regime in Fig. 6.26 and of the subsequent sequence of states can be provided by the map (6.58) assuming that the curve is now tilted on the x_n axis as shown in Fig. 6.30.

Several iterations are shown in Fig. 6.30 for a generic starting point P from within the basin of attraction of the stable fixed point A which extends to the left of x_E. The points to the right of E, for example Q, have iterates which remain trapped above the segment EM giving rise to an aperiodic behaviour (the fixed point B is unstable). It is precisely due to the presence of the minimum, M, which lies above the point E, that the iterates to the right of E do not succeed in re-entering into the basin of attraction of A. The basins of attraction of the periodic and the chaotic attractors are, however, contiguous so that it is plausible that the presence of fluctuations can shift the system from one to the other. For example, when the iterates fall into a region around M, a small additional fluctuation is sufficient to cause the next iterate to jump into the basin of attraction of A and viceversa, from the latter towards the basin of the chaotic attractor. This may induce a random oscillation between the neighbouring basins of attraction and give rise to a time record similar to that shown in Fig. 6.26. The prolonged laminar oscillations arise when the system is in state A or, in any case, in the close neighbourhood of A, under the action of further fluctuations.

A second qualitative interpretation of the time record in Fig. 6.26 may be given in terms of a tangent bifurcation in the map drawn in Fig. 6.30; it is still to be experimentally ascertained whether the intermittent behaviour in Fig. 6.26 pertains to the first or the second situation.

If the curve is deformed and moves upwards on decreasing the bifurcation parameter, the initially unstable point B will become stable with a simultaneous production on the arc NMO (see Fig. 6.31) of the sequence of states typical of unimodal one-dimensional maps.

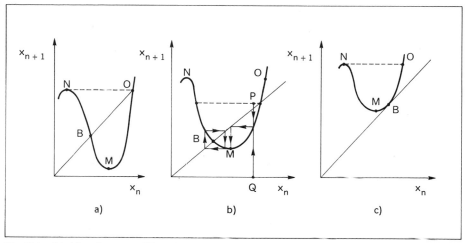

Fig. 6.31 - **(a) After the tangent bifurcation at point A in Fig. 6.30, the bisector intersects the map at the unstable points B and O. On increasing the bifurcation parameter the point B becomes stable, (b), until the latter, in its turn, vanishes after another tangent bifurcation (c).**

As the bifurcation parameter decreases, the whole curve and the arc NMO in particular undergo the changes shown in Fig. 6.31. In Fig. 6.31 (b) the construction of some iterates has been explicitly carried out with the starting point at Q. It should be noted that on turning the map upside down the arc NMO can be considered as a unimodal one-dimensional map whose deformations parallel those shown in Figs. 5.23 (c), (b) and (a) and the iterates can be obtained by means of the same construction as before with the starting point at P instead of Q. We can then conclude that the arc NMO is equivalent to the unimodal map of Fig. 5.23 and it is legitimate to expect that where B shifts from N towards M the system will go back through the U-sequence (Fig. 6.31 (a)) and then the period doubling sequence, starting from the cycles with the highest periodicity up to the cycle of period one (Fig. 6.31 (b)), corresponding to the stable fixed point B.

Some states of the U-sequence and of the "undoubling" sequence finally leading to the new limit cycle B have been effectively observed in the Belousov-Zhabotinski reaction.

A similar "transition" from a chaotic state towards a limit cycle, going backwards through the U-sequence and the period-doubling cascade, has been found in the Lorenz model for the Rayleigh-Bénard problem. On increasing the bifurcation parameter the system, initially in a chaotic state, is driven to a stable limit cycle through a sequence of states similar to that described above. This limit cycle is destabilised on further increasing the bifurcation parameter and a tangent bifurcation occurs leading once again to a new chaotic state (Manneville and Pomeau, 1980).

A similar evolution has been observed experimentally in the Belousov-Zhabotinski reaction and is predicted by the periodic map (see Figs. 6.31b and 6.31c).

On shifting the curve even further, a second tangent bifurcation occurs (see Figs. 6.31b and 6.31c). In the new intermittent regime the system shows a behaviour analogous to that shown in Fig. 6.26, the only difference being that the "laminar" regime consists of large oscillations whereas the more sporadic bursts have a smaller amplitude.

6.6 TRANSITION TO CHAOS IN THE RÖSSLER MODEL

This examination of the chaotic behaviour of deterministic systems will be concluded by presenting the results of an analytical and numerical study carried out on one of the simplest systems allowing chaotic behaviour: the Rössler system. This will provide us with the possibility of following the transition to chaos in some detail and will give us the opportunity of applying the theoretical instruments presented in Chapter 5.

Let us consider the system

$$\dot{x} = -y - z$$
$$\dot{y} = x + ay \tag{6.59}$$
$$\dot{z} = b + zx - cz$$

where a, b, and c are parameters which, for simplicity, will be assumed to be positive.

It can at once be seen that for $c < 2\sqrt{ab}$ there are no real points of equilibrium, while for $c > 2\sqrt{ab}$ there are two real and distinct points of equilibrium:

$$P_{\pm} = (-ay_{\pm}, y_{\pm}, -y_{\pm})$$

where   (6.60)

$$y_{\pm} = \frac{-c \pm \sqrt{c^2 - 4ab}}{2a}$$

The local stability properties of these points are determined by the eigenvalues of the Jacobian matrix, which are solutions of the third-degree equation

$$\lambda^3 - M\lambda^2 + N\lambda - V = 0 \tag{6.61}$$

with

$$M_{\pm} = a - (ay_{\pm} + c)$$
$$N_{\pm} = 1 - y_{\pm} - a(ay_{\pm} + c)$$
$$V_{\pm} = \mp\sqrt{c^2 - 4ab}$$

On the other hand, we know, from the general properties of third-degree equations, that

$$M = \lambda_1 + \lambda_2 + \lambda_3$$
$$N = (\lambda_1 + \lambda_2)\lambda_3 + \lambda_2\lambda_3 \tag{6.62}$$
$$V = \lambda_1\lambda_2\lambda_3$$

where λ_1, λ_2, and λ_3 are the eigenvalues of the Jacobian matrix, of which at least one is real. From Eq. (6.61) it follows that $V_{\pm} = 0$ for $c = c^* = 2\sqrt{ab}$, which implies that, in this case, at least one of the eigenvalues is zero. When $c > 2\sqrt{ab}$ then $V_- > 0$, so that the equilibrium point P_- is always unstable. On the other hand the stability of the point P_+ depends upon the values of the parameters. When a and b are fixed, the value $c^* = 2\sqrt{ab}$ corresponds to a bifurcation of the "saddle-node" type. At this point, it is to be noted that the equilibrium points P_+ and P_- may both be unstable. In fact, although on the central manifold, one is led to an equation of the same type as Eq. (5.51), which allows one stable and one unstable equilibrium point, the point P_+ is stable in the three-dimensional space only if the real parts of all three eigenvalues are positive.

In order to illustrate this fact, it can be seen that locally (i.e. in a fairly small neighbourhood around c^*) the type and the sign of the eigenvalues which do not vanish will be the same as in

$c = c^*$. But for $c = c^*$ we have

$$\lambda_1 = 0$$

$$\lambda_{2,3} = \frac{M^* \pm \sqrt{M^{*2} - 4N^*}}{2}$$

(6.63)

where

$$M^* = M(a, b, c^*) = \sqrt{a}\ (\sqrt{a} - \sqrt{b}\ )$$

$$N^* = N(a, b, c^*) = (\sqrt{a} + \sqrt{b} - a^2 \sqrt{b}\ )/\sqrt{a}$$

(6.64)

A complete study of the eigenvalues as a function of the parameters a and b would be too vast to be covered here. Thus, only a restricted region will be examined within which, however, the most significant behaviour of the system will be found.

Let us assume that $a < 1$. In this case, it is straightforward to verify that $N^* > 0$ and $M^{*2} > 4N^*$ if $b > \varphi^2$ (where $\varphi = (4 - a^2)/ a \sqrt{a}$); it follows that λ_2 and λ_3 of Eq. (6.63) are both negative real numbers. On the other hand if $b < \varphi^2$, then $M^{*2} < 4N^*$, so that λ_2 and λ_3 are complex conjugate; their real part is $\gtrless 0$ according to whether $a \gtrless b$. Let us now assume that a and b are kept constant with $a < 1$ and $b > a$, allowing the value of c to vary. It can be shown analytically that, expressing the solutions of the third-degree algebraic equation (6.61) using the Cardano formulae, a value $c_H > c^*$ exists for which the real part of a pair of complex eigenvalues vanishes. If the conditions of the Hopf theorem are satisfied, then one may expect the birth of a stable limit cycle for $c > c_H$. The numerical simulation does, in fact, show this to be the case. On the contrary, when $b < a$ there is no Hopf bifurcation. In any case, the three projections onto the xy, xz and yz planes in Fig. 6.32 show a limit cycle corresponding to the values $a = 0.2$, $b = 0.1$ and $c = 0.65$. On keeping a and b fixed and further increasing c, the limit cycle (Fig. (6.33)) undergoes a period doubling process which continues on increasing c (see Figs. 6.34 and 6.35). We may hypothesize, at this point, that any further increase in the value of c leads to the birth of a chaotic attractor beyond a certain value $c = c_\infty$. A determination of the value of c_∞ may be attempted using the Feigenbaum series (Eq. (5.65)), which rapidly converges. The estimated value $c_\infty = 4.15$ is in good agreement with the numerical simulation. Figure 6.36 shows the attractor for $c = 4.2$ and Fig. 6.37 shows the two-dimensional Poincaré map obtained by the intersection with the $y = -z$ plane for the case where $c = 4.2$. This consists of two branches between which the representative point of the system alternatively resides. It is then possible to obtain a unimodal map (Fig. 6.37b) plotting x_{n+1} vs. x_n. This map is qualitatively similar to the logistic map (see Fig. 5.22), and therefore we expect that the universal properties of the "transition to chaos", illustrated on the logistic map, will continue to hold also in the present case.

The fact that the complexity of this behaviour is also observed in a region where bifurcations do not appear to be foreseen by the Hopf theorem seems, at first, rather strange. It can be shown, however, that Hopf bifurcations do, in fact, occur even in this region if one moves along a suitable curve in three-dimensional space (a, b, c).

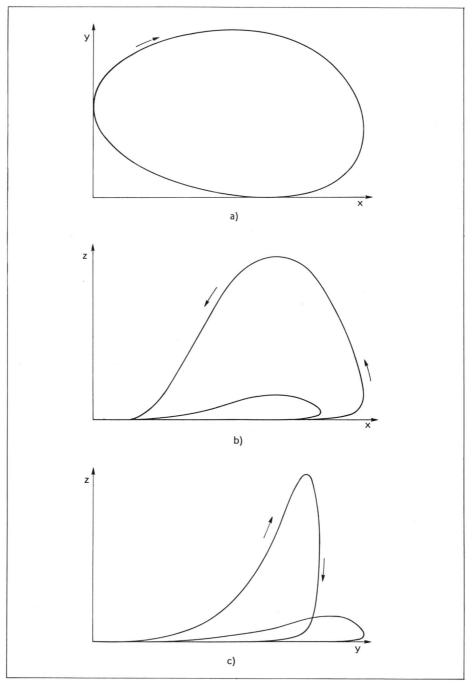

Fig. 6.32 - (a), (b), (c). Projections onto the x-y, x-z and y-z planes of the limit cycle corresponding to the parameter values $a = 0.2$, $b = 0.1$, and $c = 0.65$.

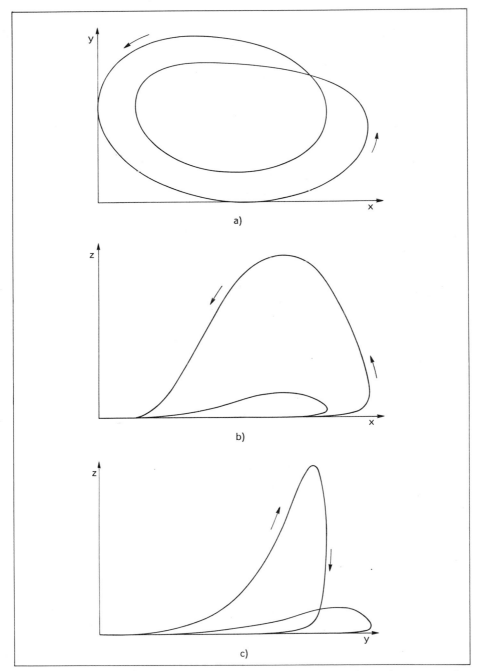

Fig. 6.33 - (a), (b), (c). *x-y, x-z* and *y-z* projections of the limit cycle corresponding to the parameter values
$a = 0.2$, $b = 0.1$ and $c = 3$; this cycle of period 2 develops from the previous one shown in Fig. 6.32
by period doubling.

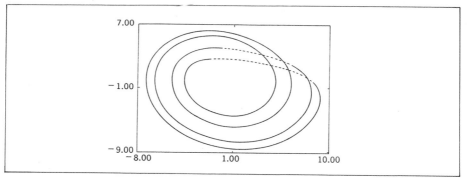

Fig. 6.34 - *x-y* projection of the limit cycle of period 4 corresponding to the parameter values $a = 0.2$, $b = 0.1$ and $c = 4$ (after Gardini, 1985).

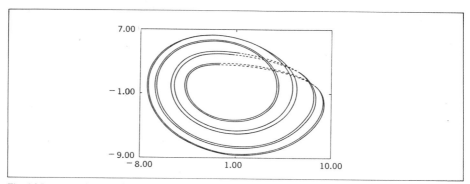

Fig. 6.35 - *x-y* projection of the limit cycle of period 8 corresponding to the parameter values $a = 0.2$, $b = 0.1$ and $c = 4.05$ (after Gardini, 1985).

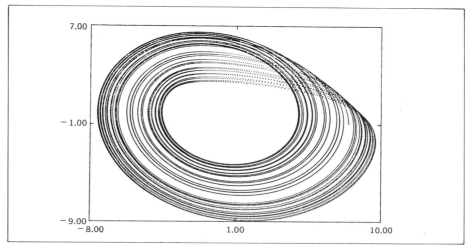

Fig. 6.36-*x-y* projection of the chaotic attractor corresponding to the parameter values $a = 0.2$, $b = 0.1$ and $c = 4.2$ (after Gardini, 1985).

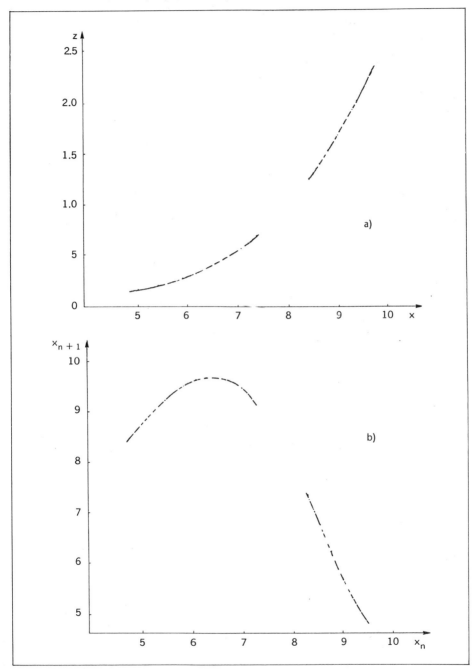

Fig. 6.37 - (a) Two-dimensional Poincaré map obtained by intersecting the trajectory shown in Fig. 6.36 with the plane $x = -z$. (b) One-dimensional map x_{n+1} vs. x_n obtained from the previous two-dimensional map (after Gardini, 1985).

6.7 THE INTERPLAY OF STOCHASTIC AND DETERMINISTIC NOISE

The interpretation of experimental data has so far been carried out using exclusively kinetic models based upon the detailed representation of chemical reaction mechanisms or upon reduced deterministic descriptions provided by the one-dimensional maps.

These theoretical models have allowed a qualitatively satisfactory description of the observed regimes, with the exception of the so-called flip-flop chaos and the intermittent regime in Fig. 6.25. The role of fluctuations in the genesis of these states leads one to think that, in such cases, the interference of stochastic noise with the deterministic behaviour can produce more than marginal effects which are worth investigating in more detail.

Some indications may arise from the study of a relatively simple system of the same type as those employed for studying the insurgence of turbulent regimes in condensed matter systems. Such a system is the damped anharmonic oscillator immersed in a double-well potential $V(x) = ax^2/2 - bx^4/4$ (analogous to that shown in Fig. 1.4) which is simultaneously subjected to an external periodic force. Precisely because of the nonlinearity of the potential, the system possesses a rather complex deterministic behaviour which has been simulated using an analogue computer with the aim of showing behaviour which would not be visible using normal perturbative techniques (the latter, in fact, only allow exploration of small amplitude oscillations and do not cover the phenomenology relating to large displacements from the equilibrium position). Over a certain range of parameters, using the period of the external force, T_d, as a bifurcation parameter, a cascade of period doubling bifurcations was found, eventually leading to chaos according to the Feigenbaum scheme. Also, as is to be expected from the type of phenomenology observed, from the Poincaré sections of the attractor one-dimensional return maps with a quadratic maximum were obtained. It should be stressed here that the forcing term renders the two-dimensional system non-autonomous and this amounts to introducing an additional degree of freedom (time). Likewise, the forced Brusselator also exhibits chaotic behaviour (see Schneider, 1985).

The chaotic regime, in its turn, is characterised by the evolution of the attractor from a structure with separate bands to a single band structure. This metamorphosis of the attractor occurs through a sequence of successive mergings in which the contiguous pairs of bands merge into single bands (a detailed study of this attractor for different values of the damping parameter is to be found in Moon and Li, 1985).

The next step was to simulate the behaviour of the system itself in the presence of an additive white noise source. The strange attractor, which describes the chaotic motion of the oscillator, was stable even when subject to fairly intense fluctuations, its major topological characteristics remaining unchanged. The most interesting effect was, however, found in the sequence of periodic states preceding chaos and, symmetrically, upon the sequence of the mergings of the attractor bands.

It was thus found that an increasing number of states, initially accessible to the deterministic system, become inaccessible to the stochastic system.

On increasing the intensity of the fluctuations, as shown schematically in Fig. 6.38, the sets of periodic and chaotic states in a continually expanding neighbourhood of $P = \infty$ can be no longer observed.

In the bifurcation diagram of the stochastic system, the periodic states are lost (starting from those with the highest periods) together with, in a symmetrical manner, the chaotic states (corresponding to the highest number of bands), enclosed within the shaded area in Fig. 6.38.

On the periodic side of this region a noise-induced chaotic regime is established due to the presence of noise; accordingly, the transition to chaos is anticipated in the sense that the chaotic regime sets in after a finite number (no longer infinite as in the deterministic case) of bifurcations. On the chaotic side of the shaded region of Fig. 6.38 the action of noise imposes an

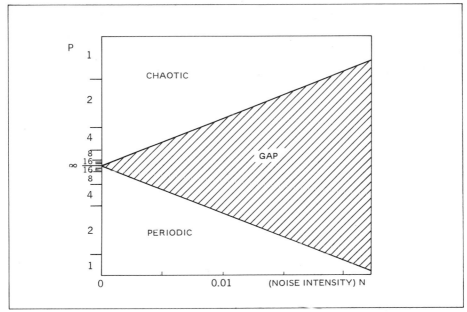

Fig. 6.38 - Diagram of the states which are inaccessible to the forced nonlinear oscillator as a function of a measure of the noise intensity. On the vertical axis the periodic states are listed according to their periodicity, while the chaotic region is divided into windows labelled according to the number of bands of the chaotic attractor (after Crutchfield and Huberman, 1980).

upper limit to the number of distinguishable bands in the chaotic attractor. From Fig. 6.38 it is clear that the maximum periodicity or the maximum number of observable attractor bands decreases on increasing the noise intensity (in the periodic range, below $P = \infty$, the parameter P on the vertical axis is defined as the ratio between the actual period of the oscillator T and the period, T_d, of the driving force; in the chaotic range, above $P = \infty$, P indicates the number of bands of the chaotic attractor).

The presence of noise implies a renormalisation of the threshold value of the bifurcation parameter for the transition to chaos. The relationship existing between the noise intensity, N, and the threshold value of the bifurcation parameter beyond which chaos is established, T_{th}, is of the following type:

$$T_{th} = T_{th}^{\infty} (1 - N^{\gamma}) \tag{6.65}$$

where T_{th}^{∞} is the threshold value of the driving period for the transition to chaos in the deterministic case, and γ is an exponent, which in this case is of the order of unity.

The study of the effects of noise upon the deterministic dynamics of the system is significantly helped if carried out using a more compact description of the deterministic system in terms of maps. For example, the analysis of the anharmonic oscillator already mentioned above may be carried out on the related one-dimensional map. In the following we shall restrict our attention to the behaviour of one-dimensional maps with a single maximum, without further investigating the problem of constructing the map from the full equations of the system. The choice of a particular map will not be so crucial as far as universal properties are concerned. In fact they are shared by all the maps belonging to the same class (in our case the class of single maximum one--dimensional maps). We can then refer back again to the logistic map (5.65) whose behaviour has already been outlined in Section 5.7:

$$\xi_{n+1} = \alpha \xi_n (1 - \xi_n) \qquad 0 < \xi_n < 1 \qquad\qquad (6.66)$$

Figure 6.39 gives the bifurcation diagram for the map represented by Eq. (6.66), from which the bifurcation sequence can be clearly seen, which, at the threshold value α_∞ leads to chaotic behaviour. Beyond α_∞ there are windows of periodic behaviour where cycles of a period which is an integral multiple of the fundamental period T occur, together with the related period doubling sequences of the $2^n K$ type (with $K = 2, 3, \ldots$; $K = 1$ corresponds to the first cascade for $\alpha < \alpha_\infty$).

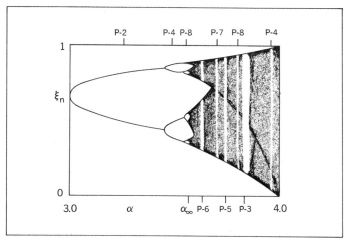

Fig. 6.39 - Bifurcation diagram for the logistic map with the bifurcation parameter α ranging over the interval $[3, 4]$ (after Crutchfield, Farmer and Huberman, 1982).

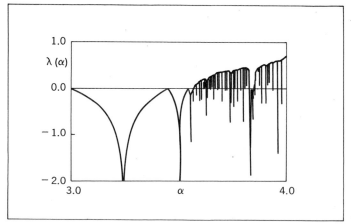

Fig. 6.40 - Behaviour of the Lyapunov exponent λ for the logistic map as a function of the bifurcation parameter α (after Crutchfield, Farmer and Huberman, 1982).

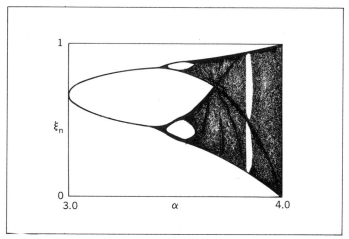

Fig. 6.41 - Bifurcation diagram for the logistic map under the effect of additive noise (after Crutchfield, Farmer and Huberman, 1982).

A more decipherable image of the bifurcation diagram given in Fig. 6.39 is shown in Fig. 6.40 which gives the behaviour of the Lyapunov exponent λ as a function of α, calculated according to Eq. (6.57).

The function $\lambda(\alpha)$ takes negative values over the interval $[3, \alpha_\infty]$ where the map predicts the emergence of periodic attractors, while the zeroes indicate the succession of the various bifurcations. The zeroes of the curve become more frequent near to α_∞, beyond which the Lyapunov exponent takes positive values indicating the formation of the chaotic attractor. Even in this region there are points with negative values of $\lambda(\alpha)$ which correspond to the windows of periodicity in the U-sequence and to the related series of period doubling bifurcations.

Let us now add a Gaussian additive noise term, p_n, with a zero average and fixed standard deviation, to Eq. (6.66):

$$\xi_{n+1} = \alpha \xi_n (1 - \xi_n) + p_n \qquad (6.67)$$

and examine the changes which occur in the system's behaviour. Figure 6.41 gives the bifurcation diagram in the presence of noise. In comparison with Fig. 6.39, it can be seen that the stochastic term in Eq. (6.67) causes the loss of the detailed structure which characterised the deterministic diagram. For example, the succession of periodic windows in Fig. 6.39 for $\alpha > \alpha_\infty$ seems to have disappeared except for the window P_3. Also, as was to be expected, the succession of sharply defined trajectories before α_∞ degenerate into a band-like structure.

This effect is particularly visible if a comparison is made between the probability functions $P(\xi)$ for the two-band attractor for $\alpha = 3.59687$ in the presence and in the absence of noise (the probability function gives the relative frequency with which the representative point visits the various points of a section of the chaotic attractor). It can be seen from Fig. 6.42 (a) that the system tends to prefer particular values of ξ, corresponding to isolated peaks in the profile of the probability function, whereas it visits the neighbourhood of each of these values much less frequently. Fig. 6.42 (b) shows that the additive fluctuation diverts the system from its preferred trajectories, delocalising it over the neighbourhoods of the most probable values of the deterministic map. The probability peaks are thus distributed over their respective neighbourhoods and the uneven irregular profile in Fig. 6.42 (a) is transformed into the smoother one in Fig. 6.42 (b).

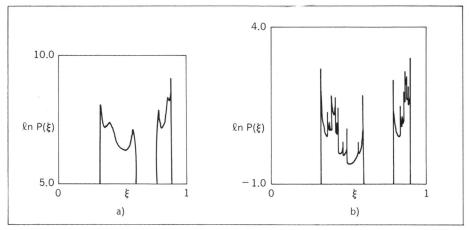

Fig. 6.42 - **Logarithmic plots of the unnormalised probability function for the chaotic attractor of the logistic map when the attractor consists of only two bands; (a) without noise and (b) with additive noise (after Crutchfield, Farmer and Huberman, 1982).**

It may happen that there are no other attractors in the immediate neighbourhood of the trajectory, upon which the fluctuations can divert the system. In this case, the representative point will again be attracted by the original orbit until the next fluctuation arrives. The resulting trajectory will then take the form of a small band centred around the deterministic trajectory.

This is the effect which is clearly visible in Fig. 6.41 on the first bifurcation of the period doubling sequence for $\alpha < \alpha_\infty$ whose "branches" are clearly separated, at least for the given noise intensity. Naturally, if the fluctuations are sufficiently intense, the point may undergo transitions between stable contiguous "branches" which will thus tend to merge into a single band.

It is to be noted that, for the same noise levels, this merging is all the more likely when the system is closer to α_∞ since the bifurcation "branches" are closer to each other.

This explains why the bifurcations of lowest periodicity are the most resistant to the action of the fluctuations. Before reaching α_∞ in Fig. 6.41, only the period 4 attractor can still be distinguished for the given noise level whereas the period 8 orbit, which could be clearly distinguished in Fig. 6.39, disappears.

Even in the chaotic portion, the disappearance of the periodic windows of the U-sequence is brought about by the degeneration of the periodic attractors into bands which may merge into a single chaotic band. Figure 6.4(e) illustrates the succeeding stages of this process on the period 3 window in the chaotic region (the P_3 window in Fig. 6.39 which can still be distinguished in Fig. 6.41). Clearly, the progressive broadening of the three peaks, which originally represented the period 3 orbit, is followed by the appearance, with increasing noise intensity, of a broad background which fills the gap between the peaks until they merge into a single chaotic band.

It is worth mentioning that the intermittent regime which arises on passing from an odd-limit cycle towards the neighbouring chaotic state has been investigated in the logistic map. We refer the reader to the paper by Hirsch and others (1982) where the transition from the P_3 window to the chaotic state for $\alpha < \alpha_3$ has been thoroughly studied. Finally, scaling laws for the Lyapunov exponent and for the average length of the laminar stage in the absence and in the presence of noise have been determined, shedding light on the synergic effect of complex deterministic behaviour and stochastic noise.

The study of the map (6.67) also leads to the identification, in the bifurcation diagram, of a gap of states which are no longer accessible to the system, analogous to that in Fig. 6.38 and

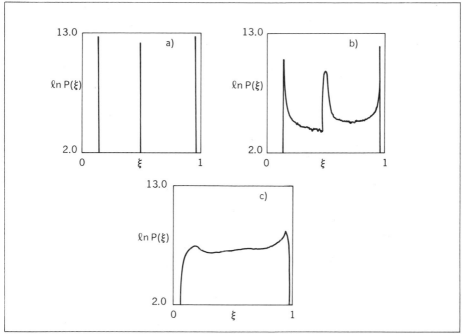

Fig. 6.43 - (a) Logarithmic plots of the unnormalised probability density at the superstable period 3 orbit for $\alpha =$ = 3.831874. On increasing the noise level, as in (b) and (c), the delta functions broaden and eventually merge into a single chaotic band (after Crutchfield, Farmer and Huberman, 1982).

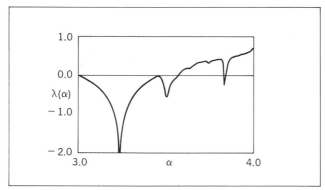

Fig. 6.44 - Behaviour of the Lyapunov exponent of the logistic map as a function of the parameter α; the noise intensity is the same as in Figs. 6.41 and 6.42 (b) (after Crutchfield, Farmer and Huberman, 1982).

which can be associated with a scaling law of the threshold for the transition to chaos identical to that in Eq. (6.65).

In the presence of noise, noisy periodic attractors and chaotic attractors are hardly distinguishable. Discriminating between them is much easier with the aid of the Lyapunov exponent, whose plot versus the bifurcation parameter is shown in Fig. 6.44.

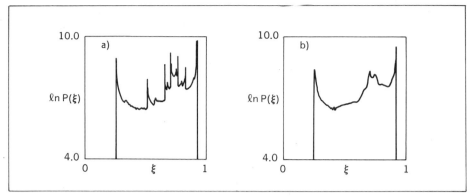

Fig. 6.45 - **Comparison between the logarithmic plot of the unnormalised probability density functions for the single band attractor at** $\alpha = 3.7$ **for a lower noise level, (a) and for a higher one, (b) (after Crutchfield, Farmer and Huberman, 1982).**

A comparison of Fig. 6.44 with Fig. 6.40 confirms the qualitative observation made from the bifurcation diagram. The disappearance of the bifurcations below α_∞ and of the periodic windows in the chaotic region is indicated by the fact that the curve remains above the axis and loses the extremely irregular behaviour characterised by the numerous peaks (pointing downwards) present in the deterministic case. Finally, the transition to chaos, hallmarked by the appearance of positive values of λ, is anticipated with respect to the noiseless case in Fig. 6.40 in close analogy with the phenomenon discussed above for the anharmonic oscillator.

Regarding the effect of noise upon the single band attractor, Fig. 6.45 shows that the essential characteristics of the profiles of the probability functions do not undergo significant changes.

This examination of the Lyapunov exponent, given in Fig. 6.46 as a function of the noise level, shows that the exponent is practically constant over that range of noise intensity and thus confirms that the "degree of disorder" of the chaotic state remains constant notwithstanding the increasing noise level.

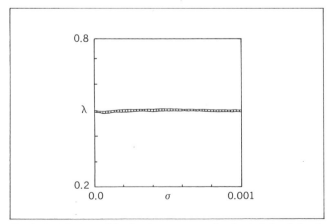

Fig. 6.46 - **Behaviour of the Lyapunov exponent for the single band chaotic attractor of Fig. 6.45 over the same range of noise levels (after Crutchfield, Farmer and Huberman, 1982).**

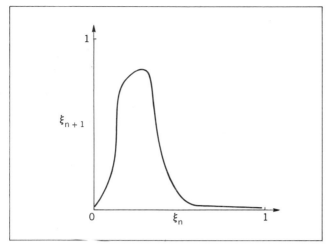

Fig. 6.47 - Behaviour of the map (6.68) (after Matsumoto and Tsuda, 1983).

Some of the conclusions drawn during the discussion of the logistic map can not, however, be indiscriminately generalised. A particularly interesting effect emerged during the study of a map proposed for the interpretation of a certain experimental bifurcation sequence in the Belousov-Zhabotinski reaction.

The map in question

$$x_{n+1} = [-a^{1/3} + 0.50607357]\exp(-x_n) + b \qquad x_n < 0.125$$
$$x_{n+1} = [a^{1/3} + 0.50607357]\exp(-x_n) + b \qquad 0.125 < x_n < 0.3 \qquad (6.68)$$
$$x_{n+1} = 0.121205692\,[10x_n\exp(-10x_n/3)]^{19} + b \qquad x_n > 0.3$$

with $a = (x_n - 0.125)$, contains an additive bifurcation parameter, b, and behaves as shown in Fig. 6.47.

If the parameter b is allowed to fluctuate, i.e. if additive fluctuations, p_n, are added to the map (6.68), a new type of interference between the deterministic and the stochastic noise arises which lends greater order to the chaotic state than it possessed previously in the absence of noise-induced order).

On examining the map for $b = 0.023288$ (the following observations can be made for various values of b), the Lyapunov exponent, at first positive, tends to take negative values with increasing noise intensity (see Fig. 6.48).

Correspondingly, as seen in Fig. 6.49, a peak emerges in the power spectrum which indicates the transition towards a "more ordered" state.

In terms of probability density of the attractor, shown in Fig. 6.50, the transition not only brings about a smoothing of the pronounced peaks in the noiseless case, but also a reduced dispersion over the interval with an increase at the edges and, in particular, the appearance of the central peak at the unstable fixed point.

The fact that the profiles of the probability density in Fig. 6.50 are roughly similar to those of a period 6 orbit with the same noise level makes the "ordering effect" of noise particularly apparent.

The main contribution to the appearance of noise-induced order comes, above all, from a mechanism analogous to that invoked previously in the case of the logistic map.

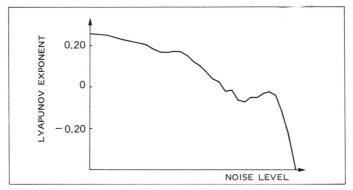

Fig. 6.48 - Dependence of the Lyapunov exponent of the map (6.68) upon noise intensity (after Matsumoto and Tsuda, 1983).

The fact is that the noise tends to modify the probability density profile of the deterministic map with a profile resulting from the average carried out over a region around each point. This "averaging" operation does not favour isolated peaks, such as the central one, in the sense that it tends to reduce them while its effect upon fairly close peaks is that of merging them into each other as observed with the bands of the attractor in the logistic map.

The transition to a more ordered state is also favoured by the characteristics of the deterministic map (6.68) which are enhanced by the randomising action of the noise. The fact is that the map (6.68) favours the grouping together of the iterates at the interval edges, due to the predominance of low gradient regions in the curve in Fig. 6.47. Fig. 6.51 shows how the main portion of the interval (shaded areas) is mapped onto the edges of the vertical interval.

If, at the same time, the additive fluctuations transform x_{n+1}, obtained from the deterministic map, into $x_{n+1} + p_n$, it is clear that there will be a greater probability of the system falling within the shaded subset (rather than within its complement with respect to the interval) and this leads, with each iteration, to a continuous focussing effect towards the edges of the interval. In the light of this mechanism, not only is the disappearance of the central peak more comprehensible, but so is the consolidation of the three peaks on the left together with the emergence of the peak at the extreme right of the interval.

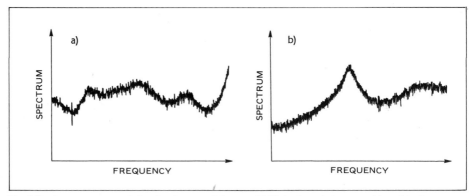

Fig. 6.49 - Power spectrum of the chaotic state of the map (6.68) in the noiseless case, (a), and in the presence of additive noise, (b) (after Matsumoto and Tsuda, 1983).

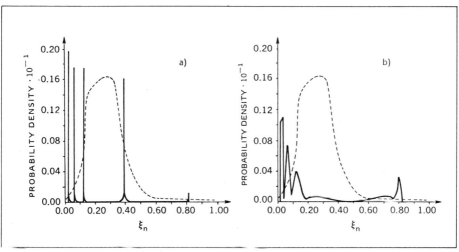

Fig. 6.50 - Comparison of the probability densities for the chaotic state in the map (6.68) (a) in the noiseless case and (b) in the presence of noise. The dotted line in the background represents the profile of the map (see Fig. 6.47) (after Matsumoto and Tsuda, 1983).

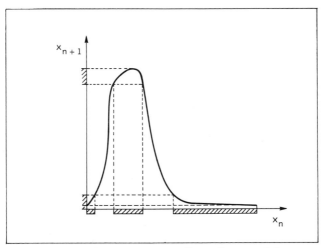

Fig. 6.51 - The map (6.68) produces an effective focussing of the iterates towards the edges of the interval. The iterates of the main part of the interval on the x_n axis (shaded region) are condensed into a much smaller region (indicated by the shaded region) on the x_{n+1} axis.

6.8 NOISE-INDUCED TRANSITIONS

In the previous Chapters we have already met multiplicative noise. Section 3.4 gives a definition of the phenomenon along with a discussion of the ambiguity in the choice of the stochastic calculus to be employed.

We have seen that multiplicative noise may arise in nonlinear systems, from the adiabatic elimination of fast degrees of freedom coupled with both order parameters and with a thermal bath through standard additive terms of the Langevin type.

Another possible cause of the appearance of multipicative noise terms is through coupling with a rapidly fluctuating external environment. The self-organisation phenomena considered so far are responses of the physical system to variations in one or more external control parameters, such as, for example, the intensity of pumping in a laser. It will now be shown that some of the systems are capable of self-organisation even in response to extremely rapid fluctuations in the control parameters.

The simplest system of this kind is probably that described by the Verhulst equation, which we have already met in Chapter 5 (Eq. (5.11)). It will be recalled that the Verhulst equation describes the growth of a population in the presence of limited resources and it has been applied, apart from animal and vegetable populations, to the description of the diffusion of technological innovations, the penetration of products in given markets, and the development and decay of energy sources. It is written here as

$$\dot{x} = hx - x^2 = -\frac{dU(x)}{dx}$$

$$U(x) = -\frac{h}{2}x^2 + \frac{1}{3}x^3 \tag{6.69}$$

For obvious physical reasons the analysis will be limited to the half-space $x \geqslant 0$.

The stability analysis has already been carried out in Section 5.2, and the conclusions may be summarised as follows. There are two equilibrium points, $x = 0$ and $x = h$. For $h > 0$, which is the physically interesting case, the first is unstable and the second stable. It is also possible to obtain the analytical solution of the Verhulst equation which is given by the logistic curve:

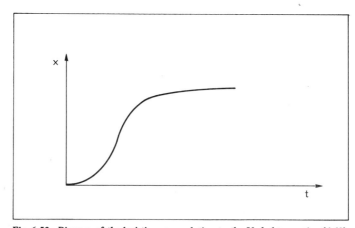

Fig. 6.52 - Diagram of the logistic curve, solution to the Verhulst equation (6.69).

$$x(t) = h x(0)/\{x(0) + [h - x(0)] e^{-ht}\} \tag{6.70}$$

The behaviour of this function is shown in Fig. 6.52.

If we were now to add a fluctuating force $A(t)$ to the r.h.s. of Eq. (6.69) we would obtain a Langevin equation. The stationary solution of the associated Fokker-Planck equation, as we know, would behave as $\exp[-U(x)/D]$, as shown in Fig. 6.53.

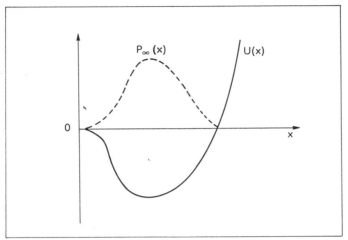

Fig. 6.53 - Profile of the potential $U(x)$ for the Verhulst equation (6.69) and plot of the asymptotic probability density in the case where a white additive force is added to the Verhulst equation.

It is to be noted that the deterministic equation (6.69) and the associated Langevin equation predict an analogous qualitative behaviour of the system; in fact the maximum of the probability density function is located at the minimum of the potential $U(x)$ (see Fig. 6.53).

In particular, it may be concluded that the system tends towards the point $x = h$. Following the treatment given by Horsthemke and Lefever we shall now see what happens if, instead of adding a fluctuating term, we assume that the coupling with the environment causes a rapid fluctuation of the coefficient h.

$$h = H + A(t)$$

$$\langle A(t) \rangle = 0 \tag{6.71}$$

$$\langle A(0) A(t) \rangle = 2 D \delta(t)$$

H is the constant part of h, to which a white noise term is to be added. The dynamic equation is then

$$dx(t) = f(x(t)) dt + \sqrt{2D}\ g(x(t)) dw(t)$$

$$f(x) = Hx - x^2 \qquad g(x) = x \tag{6.72}$$

where $w(t)$ is a standard Wiener process, that is, a process having a unitary diffusion coefficient. This stochastic process may be associated, if we adopt the Itô method, with the following Fokker-Planck equation:

$$\frac{\partial p(x,\, t)}{\partial t} = - \frac{\partial\,[(Hx - x^2)\,p(x,\, t)]}{\partial x} + D\,\frac{\partial^2\,[x^2 p(x,\, t)]}{\partial x^2} \tag{6.73}$$

According to the Stratonovich approach, however, we should write

$$\frac{\partial p(x,\, t)}{\partial t} = - \frac{\partial\,[(Hx - x^2 + Dx)\,p(x,\, t)]}{\partial x} + D\,\frac{\partial^2\,[x^2 p(x,\, t)]}{\partial x^2} \tag{6.74}$$

As we shall see, the differences associated with the choice of the appropriate stochastic calculus will not influence the conclusions deriving from our reasoning, which will be qualitatively the same in both cases. In fact, it is possible to show (see, for example, the book by Horsthemke and Lefever cited in the Bibliography) that the stationary solution of Eqs. (6.73) and (6.74) is of the following type:

$$p_\infty(x) = N g(x)^{-q}\,\exp\left[\frac{1}{D}\int_0^x \frac{f(u)}{g(u)^2}\,du\right] \tag{6.75}$$

with $q = 2$ in the case of the Itô calculus and $q = 1$ if the Stratonovich approach is used. N is a normalisation constant. The extrema of the asymptotic probability distribution are then given by the solutions of the equation obtained by putting the derivative of the previous expression equal to zero:

$$f(x) - qDg(x)\,g'(x) = 0 \tag{6.76}$$

The maxima in the asymptotic probability distribution are crucial quantities in the analysis of the behaviour of the system, as they indicate the regions in which there is the maximum probability of finding the system, after the transient has completely decayed.

In our case, Eq. (6.76) becomes

$$Hx - x^2 - qDx = 0$$

whose solutions are

$$x = 0 \tag{6.77a}$$

$$x = H - qD \tag{6.77b}$$

On studying the second derivatives of $p_\infty(x)$ it can then be shown that the abscissa of the relative maximum of the probability density takes the value (6.77b) which is physically meaningful if $H > Dq$, while, in the case $H < qD$, it takes the value (6.77a).

We can thus see what happens: assuming that we start from the limiting case of zero noise intensity, we find the probability distribution peaking at the asymptotic value of the deterministic model, $x = H$. On increasing D, the intensity of the multiplicative noise, the maximum moves towards the origin and, at the threshold value $D = H/q$, the probability distribution peaks at the origin. If we now look at the form of the asymptotic probability density for the Verhulst equation, from Eq. (6.75) we have:

$$p_\infty(x) = N x^{(H/D) - q}\,\exp\left(-\frac{x}{D}\right) \tag{6.78}$$

Thus we see that at $D = H/q$ the probability distribution has a singularity in the origin.

We thus have an abrupt transition induced by multiplicative noise (see Fig. 6.54), which in this particular case describes the extinction of the species x in the presence of fluctuations affecting the birth rate (provided that their intensity D is comparable to the birth rate average value H).

On examining other examples of such noise-induced transitions, it can be seen that the necessity of strong fluctuations, at least above a certain threshold value, is a common characteristic. The threshold value of the noise intensity required for the transition naturally depends upon the degree of coupling between the environment and the system.

It is worth mentioning that the existence of these surprising noise-induced transitions has been confirmed experimentally, for example, by analog simulations on nonlinear electric circuits as well by experiments carried out on a photosensitive chemical reaction (the Briggs-Rauscher reaction) under rapidly fluctuating illumination. For further details we refer the reader to the book by Horsthemke and Lefever and to the article by Faetti and others (see Bibliography).

In the light of what has been said above, we can briefly reconsider the example given in Section 3.5, involving a fluctuating potential barrier. In that particular context, the example was introduced to illustrate the adiabatic elimination procedure, but we can now see that it was an example of a noise-induced transition (note the similarity between Fig. 6.54 and Fig. 3.2). In Eqs. (3.52) the multiplicative interaction with the external environment is mediated by a fast variable ξ. Although the latter is initially coupled in a standard way with the environment, the hypothesis of an infinitely fast relaxation velocity ensures that the environmental fluctuations will be transferred to the variable v.

One may ask whether or not the multiplicative noise plays a role also in the case where the variables of the system interact through forces which derive from a potential, and are in contact with the external environment through standard Langevin terms. A model of this type is as follows:

$$\dot{x} = v \qquad\qquad \dot{v} = -\frac{\partial V(x, y)}{\partial x} - \gamma v + A(t)$$

$$\dot{y} = w \qquad\qquad \dot{w} = -\frac{\partial V(x, y)}{\partial y} - \lambda w + B(t)$$

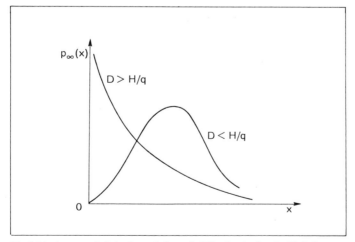

Fig. 6.54 - **Asymptotic behaviour of the probability density for the Verhulst equation (6.69) with the multiplicative noise term of Eqs. (6.71), (6.72).**

where $A(t)$ and $B(t)$ are white fluctuating forces. The model is nonlinear and this requires the presence of higher-than-second order mixed terms in V. A system of this type tends towards the asymptotic Boltzmann distribution:

$$p_\infty(x, y) \propto \exp\left[-V(x, y)/\text{constant}\right]$$

If, however, some of the variables are much faster than others (e.g. w) then, on a suitable time scale, one has the situation of multiplicative noise-induced by adiabatic elimination. It is possible that this situation, starting from initial nonequilibrium states, will have significant consequences upon the evolution kinetics of the system during the transient, even though the asymptotic distribution remains of the Boltzmann type. For a more technical discussion of this interesting possibility the reader is referred to the first five items of the bibliographical note to Section 6.8.

Appendix

PRACTICAL EXAMPLE OF AN OSCILLATING CHEMICAL REACTION

Here we present a recipe which is a slightly modified version of that published by Zaikin and Zhabotinski, but equally effective, for obtaining a system which produces interesting periodic oscillations and spatial structures.

In the following, it is assumed that all the reagents used are anhydrous and the solutions are made up with distilled water.

Sulphuric acid (2 ml) is diluted with 67 ml of water and then 5 g of potassium bromate are added, ensuring complete dissolution (the final volume of the solution will be 70 ml). 6 ml of this solution is then placed in a glass vessel together with 0.5 ml of a potassium bromide solution prepared by dissolving 1 g of potassium bromide in 10 ml of water. 1 ml of malonic acid solution (1 g in 10 ml of water) is added to the mixture, waiting for the disappearance of the slight yellow colour due to the presence of bromide. Finally, 1 ml of a standard solution (25 mM) of phenanthroline ferrous sulphate is added. The mixture after shaking is placed in a petri dish.

The evolution of the dissipative structure may be conveniently observed by illuminating the dish from below or by placing the dish on an overhead projector thereby projecting the image onto a screen. A slight effervescence (carbon dioxide) indicates that the reaction is taking place (the production of carbon dioxide also oscillates periodically, but this is rather difficult to detect).

After a short time blue spots appear in the originally red solution. Blue wavefronts branch out from these spots giving rise to a stationary interference pattern of alternate blue and red stripes. Shaking the dish leads to the disappearance of the dissipative structure which, however, is readily regenerated once the agitation ceases. If the dish is continuously stirred (using, for example, a magnetic stirrer) a periodic oscillation in the colour of the homogeneous solution occurs from red to blue and then back to red again with a characteristic period of oscillation.

Bibliography

The present bibliography is by no means intended to be a complete one. It should provide a good starting point for further reading, and furnishes the main references for a more detailed study of the topics dealt with in this book.

GENERAL TEXTS

CALLEN, H.B. (1960). *Thermodynamics*. Wiley.
An unusually clear text on a subject which is often presented in an obscure manner.

EVANS, M.W., P. GRIGOLINI, and G. PASTORI-PARRAVICINI (Eds.) (1985). *Memory Function Approach to Stochastic Problems in Condensed Matter*. Advances in Chemical Physics, Vol. 62. Wiley.
An exhaustive and very clear book on the theory of reduced stochastic models, with several applications. See also the following book:

EVANS, M.W. (Ed.) (1985). *Dynamical Processes in Condensed Matter*. Advances in Chemical Physics, Vol. 63. Wiley.

FEYNMAN, R.P., and A.R. HIBBS (1965). *Quantum Mechanics and Path Integrals*. McGraw - Hill.
One of the best physics books ever written. It contains a very clear introduction to the concept of path integral, of which Feynman is the father, along with some wonderful pages on quantum mechanics. It also deals with applications of functional integration and includes some problems on statistical mechanics.

FREHLAND, E. (Ed.) (1984). *Synergetics: from Microscopic to Macroscopic Order*. Springer - Verlag, Berlin Heidelberg.

GIHMAN, I.I, and A.V. SKOROHOD (1972). *Stochastic Differential Equations*. Springer - Verlag, Berlin Heidelberg.

GLANSDORFF, P., and I. PRIGOGINE (1971). *Thermodynamic Theory of Structure, Stability and Fluctuations*. Wiley.
A classical text of the Brussels school, emphasizing the relevance of nonequilibrium thermodynamics to the study of the stability properties of physical and chemical systems.

GUCKENHEIMER, J., and P. HOLMES (1983). *Nonlinear Oscillations, Dynamical Systems and Bifurcations of Vector Fields*. Springer - Verlag, New York.

HAKEN, H. (1978). *Synergetics. An Introduction: Nonequilibrium Phase Transitions and Self-Organisation in Physics, Chemistry and Biology*. Springer - Verlag, Berlin Heidelberg.
This is probably the introductory text which is closest to our particular treatment. The book is easily readable for most of the general aspects, but some particular questions are treated too briefly.

HAKEN, H. (1983). *Advanced Synergetics. Instability Hierarchies of Self-Organizing Systems and Devices*. Springer - Verlag, Berlin Heidelberg.

214   *Introduction to the Physics of Complex Systems*

HAKEN, H. (Ed.) (1984). *Chaos and Order in Nature.* Springer - Verlag, Berlin Heidelberg.

HUANG, K. (1963). *Statistical Mechanics.* Wiley.
An excellent text on statistical mechanics, containing an ample treatment of kinetic theory and ensemble theory.

LANDAU, L.D., and E.M. LIFSCHITZ (1959). *Statistical Physics.* Parts 1 and 2: Vols. 5 and 9 of the Course of Theoretical Physics, Pergamon Press, Oxford.
A full and authoritative text by a pioneering school in the field of phase transitions (amongst others).

LAVENDA, B.H. (1978). *Irreversible Thermodynamics.* McMillan.
Advanced text on the thermodynamics of irreversible processes which also contains an ample critical analysis of the different schools of thought on the subject. The following text, by the same author, strengthens the statistical foundations of thermodynamics.

LAVENDA, B.H. (1985). *Nonequilibrium Statistical Thermodynamics.* Wiley.

MINORSKI, N., (1974). *Nonlinear Oscillations.* Robert Krieger Publishing Co.
A justly famous treatment of nonlinear mechanics.

NICOLIS, G., and I. PRIGOGINE (1977). *Self-Organisation in Nonequilibrium Systems.* Wiley.
This text develops the premises given in the above mentioned book by Glansdorff and Prigogine. It contains a detailed exposition of the Brussels school's approach to self-organisation in dissipative structures.

PAPOULIS, G. (1965). *Probability, Random Variables and Stochastic Processes.* McGraw-Hill.
This is a particularly good text on probability theory because it combines a rigorous treatment of the subject with intuitive considerations.

POSTON, T., and I. STEWART (1978). *Catastrophe Theory and its Applications.* Pitman.

PRIGOGINE, I. (1969). *Introduction to Thermodynamics of Irreversible Processes.* Wiley.
The easiest introduction to the ideas of Prigogine.

REIF, F. (1965). *Statistical Physics.* Berkeley Physics, Vol. 5, McGraw-Hill.
A wide-ranging introduction to the subject of statistical mechanics, containing lengthy discussions on the basic principles of the statistical description of physical systems. Particularly interesting are the computer simulations of the behaviour of rigid sphere ensembles. Of notable didactic value, the text is often somewhat prolix.

RISKEN, H. (1984). *The Fokker-Planck Equation. Methods of Solution and Applications.* Springer-Verlag, Berlin Heidelberg.

SCHUSS, Z. (1980). *Theory and Applications of Stochastic Differential Equations.* Wiley, London.
A useful introduction to the subtle topics of stochastic calculus which is nonetheless accessible to the non-mathematician. The book includes several applications with particular emphasis upon the calculation of first-passage times.

SCHUSTER, H.G. (1984). *Deterministic Chaos.* Physik - Verlag, Weinheim.

STRATONOVICH, R.L. (1963, 1967). *Topics in the Theory of Random Noise.* 2 vols. Gordon and Breach.
A classical reference text as is the following.

THOM, R. (1975). *Structural Stability and Morphogenesis.* Benjamin.
The classical treatment by the founder of catastrophe theory.

TISZA, L. (1966). *Generalized Thermodynamics.* MIT Press.
A reprint of articles, complete with some unedited notes, by the extremely clear founder of this school of thought. It has the same conceptual approach as the book by Callen.

VAN KAMPEN, N.G. (1983). *Stochastic Processes in Physics and Chemistry.* North - Holland.
A notable book by one of the protagonists in the present debate within this subject. The text is well readable and is characterised by a lively polemic against the use of nonlinear Langevin-type equations which, however, we use systematically in the present work. Van Kampen's position is that only the use of Master Equations makes sense in the nonlinear case.

WAX, N. (1954). *Selected Papers on Noise and Statistical Processes.* Dover.
Extremely useful reprint of classical articles on the theory of stochastic processes, including those of Chandrasekhar, Wang and Uhlenbeck which are well worth reading even though written more than forty years ago.

SPECIFIC REFERENCES
CHAPTER 1.

CHANDRASEKHAR, S. (1943). Rev. Mod. Phys., **15** (1), 3.

KAC, M., and J. LOGAN (1979). In Montroll E.W. and J.L. Lebowitz (Eds.) , *Studies in Statistical Mechanics,* Vol. 7. North-Holland.

KRAMERS, H.A. (1940). Physica, **7**, 284.

LAVENDA, B.H. (1983). Lett. Nuovo Cimento, **37**, 200.

LAVENDA, B.H. and R. SERRA (1984). Lett. Nuovo Cimento, **41**, 417.

LAVENDA, B.H. and R. SERRA (1985). J. Math. Phys., **26**, 505.

NELSON, E. (1967). *Dynamical Theories of Brownian Motion,* Mathematical Notes. Princeton University Press.

OPPENHEIM, I., K.E. SCHULER, and G.H. WEISS (1977). *Stochastic Processes in Chemical Physics,* MIT Press.

TISZA, L., and I. MANNING (1957). Phys Rev., **105** (6), 1695.

UHLENBECK, G.E., and L.S. ORNSTEIN (1930). Phys. Rev., **36**, 823.

WANG, M.C., and G.E. UHLENBECK (1945). Rev. Mod. Phys., **17** (2 - 3), 323.

See also some of the texts cited in the general bibliography, particularly those by Haken, Van Kampen, Nicolis and Prigogine, Papoulis, Stratonovich, Gihman and Skorohod and Schuss.

CHAPTER 2

Sections 2.1, 2.2, 2.3

GOPAL, E.S.R. (1974). *Statistical Mechanics and the Properties of Matter.* Wiley.

KITTEL, C. (1958). *Elementary Statistical Physics.* Wiley.

KUBO, R. (1969). Adv. Chem. Phys., **15**, 101.

SOBELMAN, I.I., and others (1981). *Excitation of Atoms and Broadening of Spectral Lines.* Springer - Verlag, New York.

Sections 2.4, 2.5, 2.6

COLLINS, F., and G.E. KIMBALL (1949). J. Coll. Sci., **4**, 425.

NOYES, R.M. (1961). Prog. React. Kinet., **1**, 129.

NORTHRUP, S.H., F. ZARRIN, and J.A. McCAMMON (1982). J. Phys. Chem., **86**, 2314.

SOLĆ, J., and W.H. STOCKMAYER (1971). J. Chem. Phys., **54** (7), 2981.

SZABO, A., and others (1982). J. Chem. Phys., **77** (9), 4484.

VON SMOLUCHOWSKI, M. (1916). Physik. Zeitschr., **17**, 557 and 585.

VON SMOLUCHOWSKI, M. (1918). Zeitschr. f. Phys. Chem., **92**, 129.

YGUERABIDE, J. (1967). J. Chem. Phys., **47**(8), 3049.

YGUERABIDE, J., M.A. DILLON, and M. BURTON (1964). J. Chem. Phys., **40**(10), 3040.

WAITE, T.R. (1958). J. Chem. Phys., **28** (1), 103.

WILEMSKI, G., and M. FIXMAN (1973). J. Chem. Phys., **58** (9), 4009.

Section 2.7

BURSCHKA, M.A. and U.M. TITULAER (1981). J. Stat. Phys., **25**, 569.

HARRIS, S. (1981). J. Chem. Phys., **75**, 3037.

HARRIS, S. (1981). J. Chem. Phys., **75** (6), 3103.

HARRIS, S. (1982). J. Chem. Phys., **77** (2), 934.

MAYYA, Y.S., and D.C. SAHNI (1983). J. Chem. Phys., **79**, 2302.

NAQVI, K.R., K.J. MORK, and S. WALDENSTROM (1982). Phys. Rev. Lett., **49**, 304.

NAQVI, K.R., S. WALDENSTROM, and K.J. MORK (1983). J. Chem. Phys., **78**, 2710.

TITULAER, U.M. (1978). Physica, **91**A, 321.

TITULAER, U.M. (1983). J. Chem. Phys., **78**, 1004.

See also the books by Haken and by Stratonovich.

CHAPTER 3

COFFEY, W.T., M.E. EVANS, and P. GRIGOLINI (1984). *Molecular Diffusion and Spectra.* Wiley, Chap. 7, p. 215 and Chap. 8, p. 239.

COMPIANI, M., and others (1985). Chem. Phys. Lett., **114**, 503.

EVANS, M.W., and others (1982). *Molecular Dynamics.* Wiley, Chap. 9, p. 586 and Chap. 10, p. 639.

GIORDANO, M., P. GRIGOLINI, and P. MARIN (1981). Chem. Phys. Lett., **83** (3), 554.

GRIGOLINI, P. and V. ROSATO (1981). Adv. Molec. Relax. and Inter. Processes, Elsevier, p. 131.

GRIGOLINI, P., and F. MARCHESONI (1985). In EVANS, M.V., P. GRIGOLINI, and G. PASTORI-PARRAVICINI (Eds.) (1985). *Memory Function Approach to Stochastic Problems in Condensed Matter.* Advances in Chemical Physics, Vol. 62, p. 29.

MARCHESONI, F., and P. GRIGOLINI (1983). J. Chem. Phys., **78** (10), 6287.

MAZO, R. (1978). In L. GARRIDO and others (Eds.), *Stochastic Processes in Nonequilibrium Systems.* Proceedings, Sitges 1978, Springer - Verlag, Berlin Heidelberg, p. 53.

MORI, H. (1965). Prog. Theor. Phys., **33** (3), 423.

MORI, H. (1965). Prog. Theor. Phys., **34** (3), 399.

MORTENSEN. R.E. (1969). J. Stat. Phys., **1**, 271.

NORDHOLM, S., and R. ZWANZIG (1975). J. Stat. Phys., **13**, 347.

ZWANZIG, R. (1960). J. Chem. Phys., **33**, 1338.

See also the books by Gihman and Skorohod and by Stratonovich (mentioned above among the general texts) as well as:

HORSTHEMKE, W., and R. LEFEVER (1984). *Noise-Induced Transitions. Theory and Applications in Physics, Chemistry and Biology.* Springer-Verlag, Berlin Heidelberg.

CHAPTER 4

FALKOFF, D. (1958). Ann. of Phys., **4**, 325.

HASHITSUME, N. (1952). Prog. Theor. Phys., **8**, 461.

HASHITSUME, N. (1956). Prog. Theor. Phys., **15**, 369.

LAVENDA, B.H. (1977). Rivista Nuovo Cimento, **7**, 229.

LAVENDA, B.H., and E. SANTAMATO (1981). J. Math. Phys., **22**, 2926.

LAVENDA, B.H., and M. COMPIANI (1983). Lett. Nuovo Cimento, **38**, 345.

LAVENDA, B.H., and R. SERRA (1984). Lett. Nuovo Cimento, **41**, 417.

LAVENDA, B.H., and R. SERRA (1985). J. Math. Phys., **26**, 505.

MACHLUP, S., and L. ONSAGER (1953). Phys. Rev., **91**, 1512.

ONSAGER, L., and S. MACHLUP (1953). Phys. Rev., **91**, 1505.

TISZA, L., and I. MANNING (1957). Phys. Rev., **105**, 1695.

A more technical approach to nonequilibrium statistical mechanics may be found in the book by Lavenda (1985) (see General Texts).

CHAPTER 5

CHOW, S.N., and J.K. HALE (1982). *Methods of Bifurcation Theory.* Springer - Verlag, New York.

COLLET, P., and J.P. ECKMANN (1980). *Iterated Map on the Interval as Dynamical System.* Birkhauser, Boston.

GOEL, N.S., and others (1971). Rev. Mod. Phys., **43**, 231.

IOSS, G. (1979). *Bifurcation of Maps and Applications.* North-Holland, New York.

LORENZ, E.N. (1963). J. Atmospheric Sci. **20**, 448.

MARSDEN, J.E., and M. McCRACKEN (1976). *The Hopf-Bifurcation and its Applications.* Springer - Verlag, New York.

In particular, see the text by Minorski as well as that by Haken and those relating to catastrophe theory. Finally besides the book by Guckenheimer and Holmes (see General Texts), we refer to the following comprehensive collection of theoretical and experimental papers on chaos in complex systems:

CVITANOVIĆ, P. (Ed.) (1984). *Universality in Chaos.* Adam Hilger, Bristol.

CHAPTER 6

Sections 6.1, 6.2, 6.3

ARECCHI, F.T. (1981). In NICOLIS, G., G. DEWEL, and J.W. TURNER (Eds), *Order and Fluctuations in Equilibrium and Nonequilibrium Statistical Mechanics.* Wiley, New York.

HAKEN, H. (1970). *Laser Theory,* in Encyclopedia of Physics, Vol. XXV/2c, *Light and Matter* 1c, Springer - Verlag, Berlin, Heidelberg. Also reprint edition, *Laser Theory,* Springer - Verlag, Berlin Heidelberg, 1983.
HAKEN, H. (1975). Rev. Mod. Phys., **47**, 67.

Section 6.4

CRONIN, J. (1977). SIAM Rev., **19** (1), 100.
EPSTEIN, I.R., and others (1983). Scientific American, **248**, 96.
FEINBERG, M. (1980). In *Dynamics and Modelling of Reactive Systems,* Academic Press.
FIELD, R.J., and R.M. NOYES (1974). Farad. Symp. Chem. Soc., **9**, 21.
HAHN, H.S., and others (1974). Proc. Nat. Acad. Sci. U.S.A., **71**(10), 4067.
HERSCHKOWITZ - KAUFMAN, M., and G. NICOLIS (1972). J. Chem. Phys., **56**(5), 1890.
IWAMOTO, K., and M. SENO (1979). J. Chem. Phys., **70**(12), 5851.
LAVENDA, B.H., G. NICOLIS, and M. HERSCHKOWITZ-KAUFMAN (1971). J. Theor. Biol., **32**, 283.
LEFEVER, R. (1968). J. Chem. Phys., **49**(11), 4977.
PYE, K., and B. CHANCE (1966). Proc. Nat. Acad. Sci. U.S.A., **55**, 888.
SCHNEIDER, F.W. (1985). Ann. Rev. Phys. Chem., **36**, 347.
WINFREE, A.T., E.M. WINFREE, and H. SEIFERT (1985). Physica, **17D**, 109.

A clear exposition of some mathematical methods suitable for investigating the synergetic aspects of reaction-diffusion systems can be found in:

KURAMOTO, Y. (1984). *Chemical Oscillations, Waves and Turbulence.* Springer-Verlag, Berlin Heidelberg.

Section 6.5

BEHRINGER, R.P. (1985). Rev. Mod. Phys., **57**(3), 657.
BENETTIN, G., L. GALGANI and J-M. STRELCYN (1976). Phys. Rev., **14A**, 2338.
BENETTIN, G., and others (1980). Meccanica, **15**, 9.
BEN-MIZRACHI, A., I. PROCACCIA and P. GRASSBERGER (1984). Phys. Rev., **29**(2)A, 975.
BLACHER, S., and J. PERDANG (1981). Physica, **3D**, 512.
CVITANOVIĆ, P. (Ed.) (1984). *Universality in Chaos.* Adam Hilger, Bristol.
ECKMANN, J.P. (1981). Rev. Mod. Phys., **53**, 643.
ECKMANN, J.P., and D. RUELLE (1985). Rev. Mod. Phys., **57**(3), 617.
EPSTEIN, I.R. (1983). Physica, **7D**, 47.
FARMER, D., and others (1980). Ann. N.Y. Acad. Sci., **357**, 453.
FARMER, J.D., E.OTT and J.A. YORKE (1983). Physica, **7D**, 153.
FEIGENBAUM, M.J. (1983). Physica, **7D**, 16.
FIELD, R.J. (1975). J. Chem. Phys., **63**(6), 2289.
FIELD, R.J., A.KÖRÖS, and R.M. NOYES (1972). J. Am. Chem. Soc., **94**, 8649.
FIELD, R.J., and R.M. NOYES (1974). J. Chem. Phys., **60**(5), 1877.
GRASSBERGER, P., and I. PROCACCIA (1983a). Phys. Rev., **50**(5)A, 346.
GRASSBERGER, P., and I. PROCACCIA (1983b). Phys. Rev., **28**(4)A, 2591.
GRASSBERGER, P., and I. PROCACCIA (1983c). Physica, **9D**, 189.
HIRSCH, J.E., B.A. HUBERMAN and D.J. SCALAPINO (1982). Phys. Rev., **25**(1)A, 519.
HUDSON, J.L., and J.C. MANKIN (1981). J. Chem. Phys., **74**, 6171.
HUDSON, J.L., M. HART, and D. MARINKO (1979). J. Chem. Phys., **71**, 1601.
HUDSON, J.L., and others (1981). In C. VIDAL and A. PACAULT (Eds.), *Nonlinear Phenomena in Chemical Dynamics.* Springer-Verlag, Berling Heidelberg, p. 44.
MANNEVILLE, P., and Y. POMEAU (1980). Physica, **1D**, 219.
NAGASHIMA, H. (1980) J. Phys. Soc. Japan, **49**(6), 2427.
ÓRBAN, M., and I.R. EPSTEIN (1982). J. Phys. Chem., **86**, 3907.
OTT, E. (1981). Rev. Mod. Phys., **53**, 655.
PACKARD, N.H., and others (1980). Phys. Rev. Lett., **45**, 712.
PIKOVSKI, A.S. (1981). Phys. Lett., **85A**, 13.
POMEAU, Y., and P. MANNEVILLE (1980). Comm. Math. Phys., **74**, 189.
POMEAU, Y., and others (1981). J. Physique Lett., **42**, L-271.
ROUX, J-C. (1983). Physica, **7D**, 57.
ROUX, J-C., and others (1981). Physica, **2D**, 395.
ROUX, J-C., and others (1980). Phys. Lett., **77A**, 391.
ROUX, J-C., H. SIMOYI, and H.L. SWINNEY (1983). Physica, **8D**, 257.

ROUX, J-C., and H.L. SWINNEY (1981). In C. VIDAL and A. PACAULT (Eds.), *Nonlinear Phenomena in Chemical Dynamics.* Springer-Verlag, Berlin Heidelberg, p. 38.
RÖSSLER, O.E., and K. WEGMANN (1979). Nature, **271**, 89.
RUELLE, D. (1981). In C. VIDAL and A. PACAULT (Eds.), *Nonlinear Phenomena in Chemical Dynamics.* Springer-Verlag, Berlin Heidelberg, p. 30.
SCHMITZ, R.A., K.R. GRAZIANI, and J.L. HUDSON (1977). J. Chem. Phys., **67**(7), 3040.
SCHNEIDER, F.W. (1985). Ann. Rev. Phys. Chem., **36**, 347.
SHOW, R. (1981). Z. Naturforsch., **36**A, 80.
SIMOYI, R.H., A. WOLF, and H.L. SWINNEY (1982). Phys. Rev. Lett., **49**, 245.
SWINNEY, H.L. (1983). Physica, **7**D, 3.
TURNER, J.S., and others (1981). Phys. Lett., **85**A, 9.
TYSON, J.J. (1977). J. Chem. Phys., **66**(3), 905.
VIDAL, C., and others (1980). Ann. N.Y. Acad. Sci., **357**, 377.
WOLF, A., and others (1985). Physica, **16**D, 285.
WRIGHT, J. (1984). Phys. Rev., **29**A, 2924.
YOUNG, L-S. (1983). IEEE Trans. on Circuits and Systems, CAS-**30**(8), 599.

The concept of fractal dimension and its relevance to many branches of physics and mathematics has been given a lively and fascinating exposition by the father of the theory of fractals in:

MANDELBROT, B. (1977). *Fractals. Form, Chance and Dimension.* Freeman and Co., San Francisco.

Section 6.6

GARDINI, L. (1985). Nuovo Cimento, **89**D, 139.

Section 6.7

CRUTCHFIELD, J.P., J.D. FARMER, and B.A. HUBERMAN (1982). Phys. Rep., **92**, 45.
CRUTCHFIELD, J.P., and B.A. HUBERMAN (1980). Phys. Lett., **77**A, 407.
MATSUMOTO, K., and I. TSUDA (1983). J. Stat. Phys., **31**(1), 87.
MOON, F.C., and G-X. LI (1985). Physica, **17**D, 99.

Section 6.8

COMPIANI, M., and others (1985). Chem. Phys. Lett., **114**(5, 6), 513.
FAETTI, S., and others (1984). Phys. Rev., **30**(6)A, 3252.
FAETTI, S., and others (1985). In EVANS, M.V., P. GRIGOLINI, and G. PASTORI-PARRAVICINI (Eds.), *Memory Function Approach to Stochastic Problems in Condensed Matter.* Advances in Chemical Physics, Vol. 62, p. 445.
FONSECA, T., and others (1983). J. Chem. Phys., **79**, 3320.
FONSECA, T., and others (1984). J. Chem. Phys., **80**, 1826.
FONSECA, T., P. GRIGOLINI, and D. PAREO (1985). J. Chem. Phys., **83**(3), 1039.
LINDENBERG, K., V. SESHADRI, and B.J. WEST (1980). Phys. Rev., A **22**(5), 2171.
SCHENZLE, A., and H. BRAND (1979). Phys. Rev., A **20**, 1628.
WIESENFELD, K.A., and E. KNOBLOCH (1982). Phys. Rev., A **26**(5). 2946.

See also the book by Nicolis and Prigogine and references therein regarding the Brusselator.
Noise-induced transitions are thoroughly studied in the book by Horsthemke and Lefever mentioned above in the bibliography of Chapter 3.

APPENDIX

ZAIKIN, A.N., and A.M. ZHABOTINSKI (1970). Nature, **225**, 535.
WINFREE, A.T. (1972). Science, **175**, 634.

Subject Index